Lecture Tutorials
for Introductory Geoscience

Second Edition

Karen M. Kortz
Jessica J. Smay

W. H. Freeman and Company
New York

Printed in the United States of America

ISBN-10: 1-4641-0105-1

ISBN-13: 978-1-4641-0105-2

First printing

W. H. Freeman and Company
41 Madison Avenue
New York, NY 10010
Houndmills, Basingstoke
RG21 6XS England
www.whfreeman.com

Plate Tectonics and Earth's Interior

Rocks and Their Formation

Geologic Landforms and Processes

Climate Change

Historical Geology

General

PREFACE

What are Lecture Tutorials?[1]

Each Lecture Tutorial is a short worksheet that students complete in class, making the lecture more interactive. Research indicates that students learn more when they are actively engaged while learning, and several studies indicate that students who use Lecture Tutorials in the classroom retain more knowledge than students who only listen to a lecture on the same material. After a brief lecture on the subject, students work in small groups to complete the Lecture Tutorial worksheets.

The Lecture Tutorials are designed to address misconceptions and other topics with which students have difficulties. They create an environment where students confront their misconceptions and, through well-designed questioning, guide students to a more scientific way of thinking. This careful design makes Lecture Tutorials unique among most other activities used in the classroom.

By posing questions of increasing conceptual difficulty to the students, Lecture Tutorials help students construct correct scientific ideas. The first questions help the students think about what they do and do not know. The Lecture Tutorial then guides the students by asking them questions focused on underdeveloped or misunderstood concepts and slowly steps them through thinking about more difficult questions. The final questions on the Lecture Tutorial help to indicate whether the students understand the material.

Lecture Tutorials can be used in any size classroom. Students should speak with each other and teach each other while the instructor acts as a facilitator. The questions on the Lecture Tutorials require no technology and are written so the conceptual steps for each question are manageable.

Guidelines for Use

Step-By-Step Implementation for the Instructor

1. Lecture on the material as usual. You can also provide an introduction of the background information that students need to know before beginning the Lecture Tutorial.

2. Optional: Pose a well-designed, multiple-choice question for you and the students to gauge their understanding of the material.

3. Have the students split into groups of 2 or 3 and work on the Lecture Tutorial. Walk around the room and answer their questions. Lecture Tutorials take 10–15 minutes for most students to complete.

4. Review some of the main points of the Lecture Tutorial.

5. Optional: Pose a new multiple-choice question to check if the students have the expected understanding of the information.

6. Continue with the lecture.

---

[1] Revised from On the Cutting Edge—Teaching Methods—Lecture Tutorials (http://serc.carleton.edu/ NAGTWorkshops/teaching_methods/lecture_tutorials/index.html).

Directions for the Student

You are using Lecture Tutorials in your class because they help to improve your understanding of the material. They require you to actively think through questions instead of listening passively to the lecture. Lecture Tutorials allow you to gauge how well you understand the material and ask any necessary questions. They also address different learning styles, so you can use your strengths while learning the material. Surveys have shown that an overwhelming majority of students feel that Lecture Tutorials are a useful part of their learning experience.

However, in order for you to get maximum benefits from the Lecture Tutorials, you need to put effort into completing them. Think about the answers as you are working through them, and be sure to write down your logic. Nothing is more frustrating than reviewing the Lecture Tutorials and not remembering how you solved the problems! You will be asked to work in groups to complete the Lecture Tutorials. Take advantage of working with your fellow students by both learning from them and teaching them.

New to the Second Edition

The second edition of *Lecture Tutorials for Introductory Geoscience* contains nearly double the number of Lecture Tutorial worksheets from the first edition, covering a wider variety of topics. In particular, Lecture Tutorial worksheets on the topics of geologic processes, climate change, and historical geology were added to and expanded upon. In addition to the new Lecture Tutorial worksheets, the original ones from the first edition were reformatted and revised to improve their clarity and ease of use.

Acknowledgments

We wish to credit Scott Clark for his original research on misconceptions addressed in parts of eight Lecture Tutorials (Tectonic Plates and Boundaries, Subduction Features, Movement at Convergent Boundaries, Plate Boundaries in Oceans, Melting Rocks and Plate Tectonics, Transform Boundaries in Oceans, Outer Layers of Earth, The Mantle). Some of the figures and lines of questioning in these Lecture Tutorials are adapted from his published and unpublished research instruments. Initial findings were presented in "Alternative Conceptions of Plate Tectonics held by Non-Science Undergraduates" by Scott K. Clark, Julie Libarkin, Karen M. Kortz, and Sara Jordan, published in the *Journal of Geoscience Education* (2011).

We would like to thank Ed Prather and Timothy F. Slater for their guidance and encouragement. We learned of the Lecture Tutorial teaching method through the Lecture Tutorials for Introductory Astronomy they co-wrote and promoted, and after discussions with them, we created our own geoscience Lecture Tutorials.

We appreciate the enthusiasm and hard work done by Anthony Palmiotto, Anthony Petrites, Jodi Isman, Christine Buese, Amy Thorne, Scott Guile, Randi Rossignol, Clancy Marshall, Janet Bidwell, and the rest of the team at W. H. Freeman.

Each of these Lecture Tutorials has gone though many revisions, and we would like to thank all of the students who gave us feedback and comments on the previous versions of these Lecture Tutorials.

We also wish to thank Brian and Greg. Their understanding, advice, and support made this work possible.

Partial support for this work was provided by the National Science Foundation's Course, Curriculum, and Laboratory Improvement (CCLI) program under Award No. 0837185. Any opinions, findings, and conclusions or recommendations expressed in this material are those of the authors and do not necessarily reflect the views of the National Science Foundation.

PLATE TECTONICS AND EARTH'S INTERIOR

TECTONIC PLATES AND BOUNDARIES

● Below is a map showing tectonic plate boundaries. The locations where two plates meet (plate boundaries) are labeled.

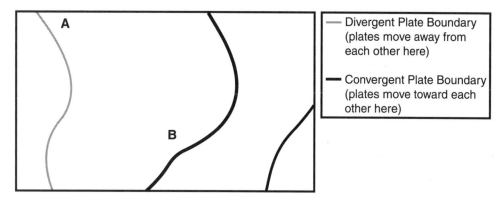

1. How many individual tectonic plate boundaries are in the diagram?

   1   2   3   4   5   6   7

2. How many tectonic plates are in the diagram?

   1   2   3   4   5   6   7

3. Are A and B on the same tectonic plate?   Yes   No

● Below is a map of the same tectonic plate boundaries. However, this map also shows the location of ocean and land.

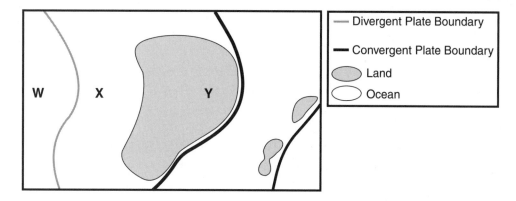

4. How many tectonic plates are in the diagram?   1   2   3   4   5   6   7

5. Why should your answers to Questions 2 and 4 match?

6. Are W and X on the same tectonic plate?   Yes   No

7. Are X and Y on the same tectonic plate?   Yes   No

8. Three students are discussing the tectonic plate on which X is located.

    Student 1: *X and Y are both on the same plate because there is no boundary between them, but X is on a different plate than W because there is a divergent boundary between them.*

    Student 2: *X and Y are on different plates because X is in the ocean and Y is on a continent. X is on the same plate as W because they are both in the ocean.*

    Student 3: *X is on a different plate than both Y and W. There is a divergent plate boundary separating X and W and an edge of a continent separating X and Y.*

    With which student do you agree? Why?

9. How is a tectonic plate <u>boundary</u> different than a tectonic plate?

10. Explain why a continent is different than a tectonic plate.

SEAFLOOR AGES

Part 1: Divergent Boundary

The diagram is of a divergent boundary with arrows showing the direction in which the plates are moving.

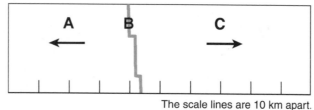

The scale lines are 10 km apart.

1. What is the age of the rocks at Location B?

    0 years old   1 million years old   2 million years old   4 million years old

2. If each plate is moving at a rate of 10 km per 1 million years, roughly how long did it take for Rock A to reach its current location?

    0 years   1 million years   2 million years   4 million years

3. What is the age of the rock at Location C?

    0 years old   1 million years old   2 million years old   4 million years old

4. Why should your answer to Question 3 be twice your answer to Question 2? Revise your answers if necessary.

5. Where is the oldest ocean crust found?   A   B   C

6. A map of the Atlantic Ocean is shown to the right. There is a divergent plate boundary running north-south in the middle of the ocean. Make a prediction: where are the oldest rocks in the Atlantic Ocean found?

    D   E

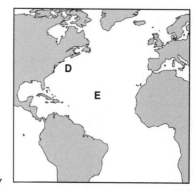

7. Two students are debating about the relative ages of the rocks that make up the crust in the Atlantic Ocean.

    Student 1: *The oldest rocks are located at E because it is the farthest from a continent. The rocks would take a really long time to get to the middle of the ocean.*

    Student 2: *But this ocean has a divergent boundary in the center. This means that rocks at E are really young. D is farthest from the divergent boundary, so that's where the oldest rocks are.*

    With which student do you agree? Why?

Part 2: The Atlantic Ocean

Examine the map of the ages of the seafloor in the Atlantic Ocean. Continents are white and outlined in gray.

8. Does the pattern of ages match your answer to Question 6? Revise your answer if necessary.

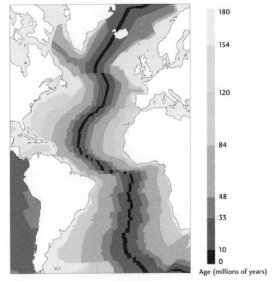

9. Draw a line along the divergent boundary.

10. What is the age of the oldest rocks in the Atlantic Ocean?

11. Approximately how long ago did the Atlantic Ocean begin to form?

_____

Map of the ages of the seafloor in the Atlantic Ocean.

12. Why should your answers to Questions 10 and 11 match? Revise your answers if necessary.

13. You are reading a proposal requesting money to search for evidence of a crater that caused a mass extinction on Earth around 250 million years ago. The team is proposing to search a poorly explored area of the floor of the Atlantic Ocean between South America and northern Africa. Would you fund this project? Use the ages of the seafloor to support your answer.

Compare your answer to the last question with the answers of other groups.

THE AGE OF THE CRUST

The cross section below shows the oceanic lithosphere (thinner, more dense) and continental lithosphere (thicker, less dense). The ocean surface is indicated by a dashed line.

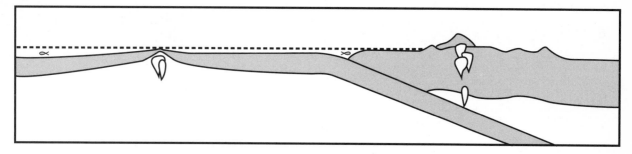

1. Label the divergent plate boundary and the convergent plate boundary.

2. What type of lithosphere is created new at the divergent plate boundary?

   oceanic lithosphere   continental lithosphere

3. What happens to the old ocean lithosphere at the convergent plate boundary?

   It is pushed underneath the surface.   It may change but stays at the surface.

4. What happens to the old continental lithosphere at the convergent plate boundary?

   It is pushed underneath the surface.   It may change but stays at the surface.

5. Based on what happens to the old ocean lithosphere, is there any very, very old ocean lithosphere remaining on Earth?

   Yes   No

6. Based on what happens to the old continental lithosphere, is there any very, very old continental lithosphere remaining on Earth?

   Yes   No

7. Label on the diagram where the oldest rock is at the surface.

8. Which statement correctly describes the relative age of continental and oceanic lithosphere?

   Continental rock is generally much older than oceanic rock.

   Oceanic rock is generally much older than continental rock.

9. The oldest continental rock is approximately 4,000 million (4 billion) years old, while the oldest ocean rock is approximately 180 million years old. What happened to the ocean lithosphere that made up the ocean seafloor 4,000 million (4 billion) years ago?

10. Two students are discussing why rocks are older on continents.

    Student 1: *Continental lithosphere is much farther from divergent plate boundaries than is oceanic lithosphere. The farther you move away from a divergent plate boundary, the older the rock gets.*

    Student 2: *That only applies to the age of the ocean rock. Old ocean lithosphere is subducted and destroyed, but continental lithosphere is not destroyed by subduction, which is why old continental rock is still around.*

    With which student do you agree? Why?

You are reading two proposals requesting money to search for the oldest rocks on Earth.

    Proposal 1: Search unexplored ocean floor beneath the ice caps in the Arctic.
    Proposal 2: Search the middle of the Australian continent.

11. Which proposal would you fund? Support your answer by explaining what happens to oceanic and continental lithosphere at tectonic plate boundaries.

12. The gray lines in the map below are divergent plate boundaries. At which location on the map is the crust older? Explain your answer.

    A: Atlantic Ocean   B: Africa

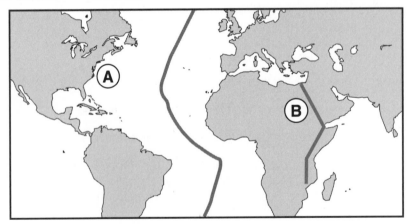

DIVERGENT BOUNDARY FEATURES

Part 1: Features at Divergent Boundaries

Below is a cross section of a divergent plate boundary. The ocean surface is indicated by a dashed line.

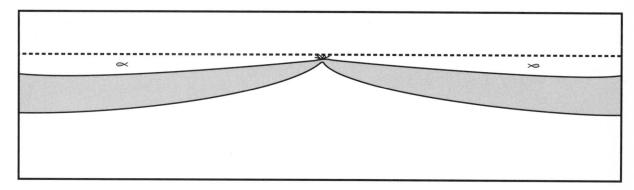

1. Draw a vertical line across the divergent plate boundary on the diagram.

2. On each plate, draw one arrow showing the direction it is moving.

3. Draw one arrow in the mantle below the lithosphere (colored gray) showing the direction the mantle is moving.

4. Circle the two features that can be found at divergent plate boundaries.

      deep ocean trench   ocean ridge   mountain range   subducting plate   lava

5. Label the cross section above with your circled features.

6. Three students are discussing one of the features you find at a divergent plate boundary.

    Student 1: *There is a ridge because the two plates are being pushed up by the hot mantle where it rises to fill in the gap as the plates spread apart.*

    Student 2: *There is a ridge because the two plates are pushing together, and they run into each other causing them to get pushed up like two slabs of rubber.*

    Student 3: *There is a trench because the two plates are pulling apart, leaving behind a large, very deep crack between them called a trench.*

    With which student do you agree? Why?

7. The map to the right shows the Atlantic Ocean, and there is a divergent boundary in the center of the ocean. Is the middle of the ocean (at the star) the deepest point? Explain your answer.

Part 2: Number of Plates

The diagram shows a divergent plate boundary.

8. How many tectonic plate boundaries are shown in the diagram? _____

9. How many tectonic plates are shown in the diagram? _____

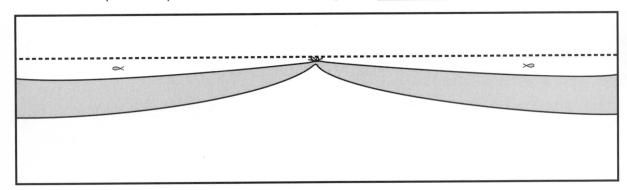

10. Two students are discussing how many plates there are in this cross section of a divergent plate boundary.

   Student 1: *There are two plates that are moving away from each other. Their edges meet at the divergent plate boundary, so there is one plate boundary.*

   Student 2: *Both sides of the boundary are ocean plates, so that means there is only one plate here, but it is stretching. I think there is one plate and one plate boundary.*

   With which student do you agree? Why?

11. How many tectonic plates are in the diagram below?_____

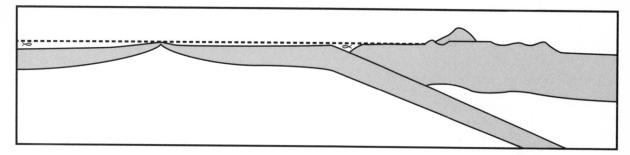

12. Label the tectonic plates you identified as 1, 2, and 3.

TRANSFORM BOUNDARIES IN OCEANS

The mid-ocean ridge on the seafloor in the Atlantic Ocean is shown below, with the divergent plate boundary shown by a thick gray. Ocean ridges are made up of a combination of divergent and transform boundaries.

1. On the image below, circle the locations of the two transform boundaries separating the divergent boundaries.

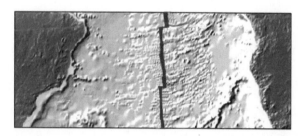

The diagram below shows a portion of an ocean ridge. The thick gray lines are divergent plate boundaries, and the thin black line is a transform plate boundary. The arrows show the overall motion of the two plates, one on either side of the plate boundary.

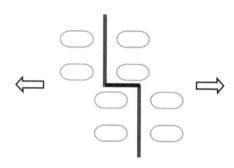

2. How many tectonic plates are in the diagram? _____

3. How many different types of plate boundaries are in the diagram? _____

4. Keeping in mind that each plate moves as a whole, draw arrows in each oval showing the direction of plate motion.

5. Are the divergent plate boundaries in the diagram above moving away from each other or do they stay in the same place? Explain your answer.

The diagrams below show a portion of an ocean ridge with two possible motions along the transform boundary. The thick gray lines represent divergent plate boundaries, and the thin black line is a transform boundary.

6. Which diagram shows the correct direction of motion at the transform boundary?   A   B

7. Check the arrows you drew on the diagram on the previous page and compare them to your answer for Question 6. Change your answer for Question 6 if necessary.

8. Two students are discussing the direction of motion at the transform boundary.

    Student 1: *I think A shows the correct motion. The arrows match the direction of motion for the rest of the plate.*

    Student 2: *I think B shows the correct motion. The arrows show the direction the divergent boundary was moved by the fault.*

    With which student do you agree? Why?

9. Is it possible for the transform plate boundary itself to extend past the divergent plate boundary as shown below? Use arrows showing plate motion on the diagram to help explain your answer.

10. Transform boundaries sometimes appear to extend past the divergent boundary along ocean ridges (see image to the right). Explain what may cause them to look this way, even though no transform movement is occurring.

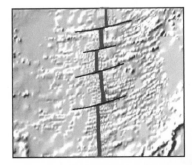

Lecture Tutorials for Introductory Geoscience

SUBDUCTION FEATURES

Part 1: Features at Convergent Plate Boundaries with Subduction

The cross section below shows a subduction zone at an ocean-continent convergent boundary. The ocean surface is indicated by a dashed line.

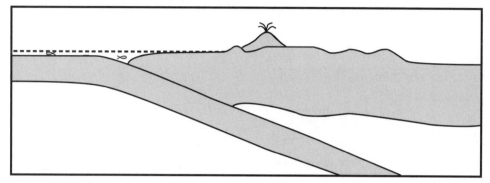

1. Draw two arrows on each plate showing which way the plates are moving relative to each other.

2. On the diagram, label features that geologists could use to identify this plate boundary.

3. For the features listed below, briefly explain how they formed.

    volcanoes and mountains:

    ocean trench:

4. Subduction of an ocean plate takes many millions of years. If you were examining a map, what would you look for to indicate that subduction is happening, even if you cannot watch?

Part 2: Sketching a Cross Section of Converging Plates with Subduction

5. Sketch one scenario that might occur when two ocean plates move toward each other. Label the trench and volcanoes.

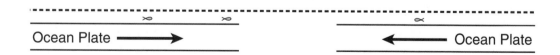

Ocean Plate ⟶   ⟵ Ocean Plate

Lecture Tutorials for Introductory Geoscience

Part 3: Plate Boundary Location

The cross section below shows a subduction zone at an ocean-continent convergent boundary.

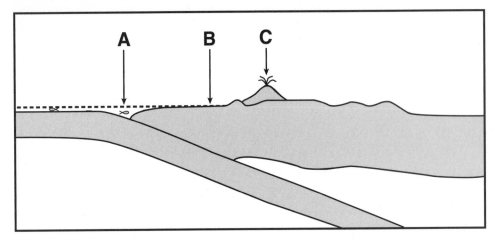

6. On the diagram, label each of the following features at the corresponding arrow:

    • volcanoes and mountains

    • coastline

    • ocean trench

7. Three students are discussing which arrow points to the location on the surface of the convergent plate boundary.

    Student 1: *I think that Arrow A points to the plate boundary because that is where one plate meets the other at the surface.*

    Student 2: *I think that Arrow B points to the plate boundary because that is where the ocean turns into continent.*

    Student 3: *I think that Arrow C points to the plate boundary because that is where the geologic action is happening.*

    With which student do you agree? Why?

8. You are given a map of an area with a subduction zone. Explain what feature you would use on the map to determine the exact location of the plate boundary.

MOVEMENT AT CONVERGENT PLATE BOUNDARIES

Part 1: Movement Shown on a Cross Section

The cross sections below show subduction zones at an ocean-continent convergent plate boundary. We will examine the boundary relative to the location of the trench. In other words, assume the location of the trench does not change, but other things might move around it.

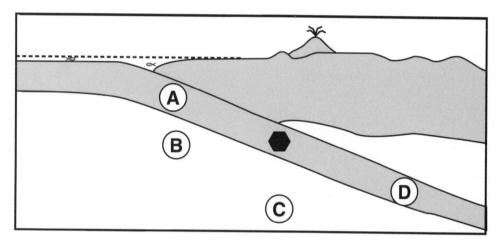

1. Draw two arrows on each plate indicating the relative direction the plates are moving.

2. Where was ⬢ in the past?   A   B   C   D   about the same place

3. Where will ⬢ be in the future?   A   B   C   D   about the same place

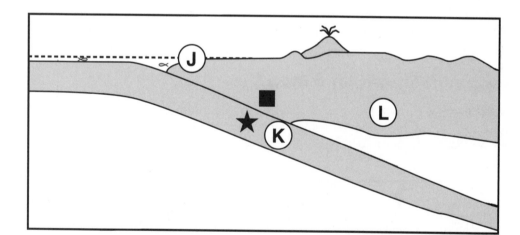

4. Where was ■ in the past?   J   K   L   about the same place

5. Where will ■ be in the future?   J   K   L   about the same place

Lecture Tutorials for Introductory Geoscience

6. Two students are discussing how the ■ on the continental plate will move over time relative to the trench.

   Student 1: *I think that the square will stay in about the same place relative to the trench. The continental plate is scrunched a little, but it isn't destroyed like the ocean plate.*

   Student 2: *But it's a convergent boundary, and the plates are moving together. Because I can draw arrows showing the plates moving together, that means that the square is moving toward the ocean plate, away from the volcano and closer to the trench.*

   With which student do you agree? Why?

7. In 50 million years, will ■ and ★ be relatively close to each other, as they are now?

   Yes　　　No

   Explain your answer.

Part 2: Movement Shown on a Map View

The map below shows the surface features at a subduction zone at an ocean-continent convergent plate boundary. We will examine the movement of features relative to the location of the trench. In other words, the general location of the trench does not change, but other things might move around it.

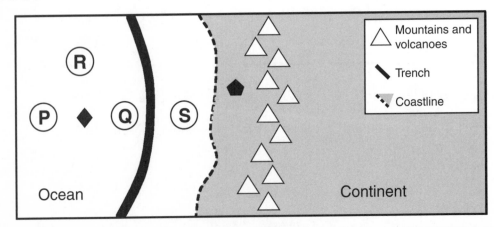

8. Draw a line along the plate boundary.

9. Draw two arrows on each plate indicating the relative direction the plates are moving.

10. Where was ◆ in the past?   P   Q   R   S   about the same place

11. Where will ◆ be in the future? P   Q   R   S   about the same place

12. At what location will ◆ disappear from view?   trench   coastline   mountains/volcanoes

13. Assuming that both ◆ and ⬢ are locations on the surface of Earth, will they ever meet?

    Explain your answer.

EFFECTS OF SUBDUCTION ANGLE

Below are diagrams of a convergent plate boundary showing ocean lithosphere subducting at different angles beneath continental lithosphere. The dashed line indicates the depth at which melting occurs as a result of subduction.

Regular Subduction:

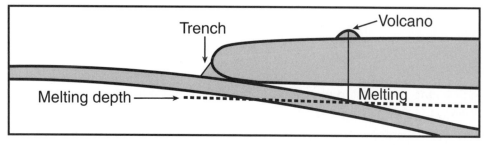

1. Describe the location where the volcano forms in terms of the subducting plate and melting depth.

2. Draw and label the locations of the trench, melting, and volcano on the two diagrams below.

High Angle Subduction:

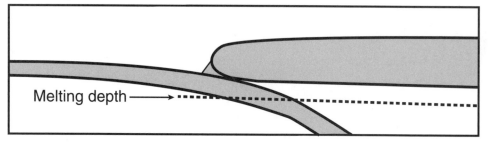

Low Angle Subduction:

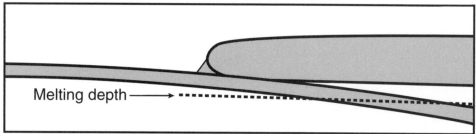

3. How does the angle of subduction influence the location of volcanoes compared to the location of the trench? Explain why the location of volcanoes changes.

The following lists the progression of volcanism and mountain-building through the Mesozoic into the Cenozoic periods:

1st Pacific coastal areas

2nd Nevada and Idaho

3rd Montana

4th New Mexico, Colorado, Wyoming

4. How did the subduction angle change during the Mesozoic?

Lecture Tutorials for Introductory Geoscience

PLATE BOUNDARIES IN OCEANS

Part 1: Features

The cross section below shows the tectonic plates beneath an ocean and a nearby continent. The ocean surface is indicated by a dashed line.

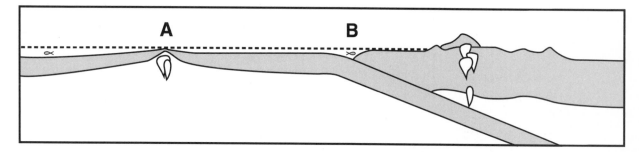

1. Draw arrows to show the directions the plates are moving near Locations A and B.

2. Two students are discussing the directions the plates are moving.

   Student 1: *I drew the arrows moving apart at Location A because magma is pushing the plates up as it rises to fill in the gap as the plates move apart.*

   Student 2: *I drew the arrows moving together at Location A because as the plates are moving toward each other, they get pushed up.*

   With which student do you agree? Why?

3. What seafloor feature is found at Location A?   ridge   trench   abyssal plain   island

4. What seafloor feature is found at Location B?   ridge   trench   abyssal plain   island

5. What type of plate boundary is at Location A?   divergent   convergent   transform

6. What type of plate boundary is at Location B?   divergent   convergent   transform

7. Check that your answers for Questions 5 and 6 match the arrows you drew at Locations A and B in Question 1.

8. If you find a plate boundary in the middle of an ocean away from the edge, what type of plate boundary is it most likely to be?

   divergent   convergent with subduction   convergent without subduction

9. If you find a plate boundary along the edge of an ocean next to a continent, what type of plate boundary is it most likely to be?

   divergent   convergent with subduction   convergent without subduction

Below is a cross section of the ocean floor and nearby land showing the surface features. The ocean surface is indicated by a dashed line.

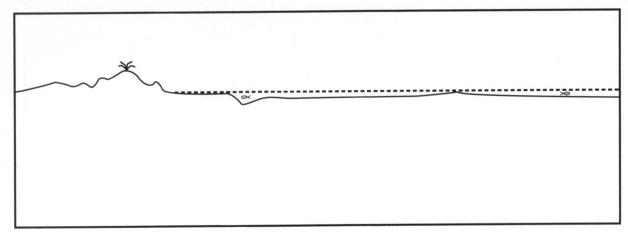

10. Label the ocean ridge.

11. Label the ocean trench.

12. Label the mountains/volcanoes on land.

13. Label the divergent boundary.

14. Label the convergent boundary.

15. Draw what the plates are doing beneath the surface to produce the surface features.

HISTORY OF AN OCEAN

Below are cross sections of an ocean at different stages of its life cycle, from pre-ocean to post-ocean. The ocean surface is indicated by a dashed line.

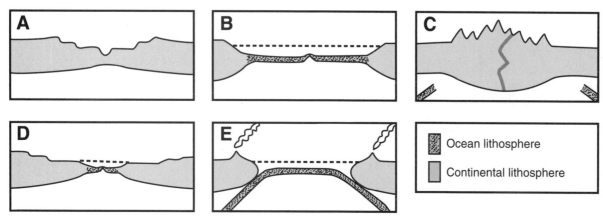

1. Describe the one plate boundary shown in B. Is the ocean growing or shrinking?

2. Describe the two plate boundaries shown in E. Is the ocean growing or shrinking?

3. Arrange the five diagrams in sequential order from pre-ocean to post-ocean. Write your order below.

   Youngest _____ _____ _____ _____ _____ Oldest

4. Two students are debating the order of the stages of an ocean.

   Student 1: *I put the cross sections in order from the smallest ocean to the biggest: C-A-D-E-B. I did this because the ocean is growing bigger over time.*

   Student 2: *But you need to look at the plate boundaries for clues. E has subduction happening, so that means the ocean is shrinking, and that means it was once bigger.*

   With which student do you agree? Why?

Lecture Tutorials for Introductory Geoscience

People often think about the age of something by watching as it changes over time (e.g., pictures of you as a baby, child, and adult). However, this technique does not work in geology because we cannot watch something change over millions of years. One way to solve this problem as it relates to oceans is to look at the oceans and put them in order of how old they are today (e.g., compare a toddler in one city to a teen in another city).

5. Match the diagrams from the previous page, which illustrate the stages of an ocean, to the five locations listed below. Place the letter next to the stage of ocean formation.

_____ East African Rift

_____ Atlantic Ocean

_____ Himalayas

_____ Red Sea

_____ Pacific Ocean

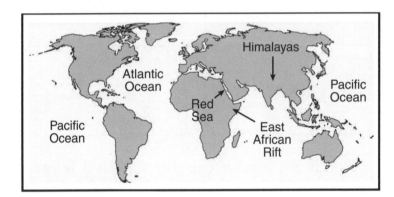

6. Take your answers to Questions 3 and 5 and place the locations on Earth in order from the youngest to the oldest in the life cycle of an ocean.

Youngest _____ _____ _____

_____ _____ Oldest

7. Two students are debating the locations on Earth and their order in the stages of an ocean.

Student 1: *I think the Himalayas are the very end of an ocean. They formed as a result of a convergent boundary, so the plates are coming together, which means the ocean that was once there completely shrunk away into nothing.*

Student 2: *No. I think the Himalayas are the very beginning of an ocean because they are the highest, and then the crust would get thinner and thinner until it becomes an ocean, and the ocean would then grow bigger.*

With which student do you agree? Why?

8. Revise the order of your ocean stages in Questions 3 and 6 if necessary.

9. Predict what the Red Sea will look like in 100 million years.

10. Predict what the East Coast of the United States could look like in 100 million years.

FEATURES ON THE OCEAN FLOOR

The ocean floor is primarily oceanic crust, but it also includes the edges of the continent (the continental margins). Although once thought to be featureless plains, there are many features that have been discovered on the ocean floor.

1. The cross section below shows the tectonic plates making up an ocean floor and two continents. The ocean surface is indicated by a dashed line. On the cross section, draw arrows to label the following ocean floor features:

| Two continental shelves | Abyssal plain (choose one) | Island | Ocean ridge | Ocean trench |

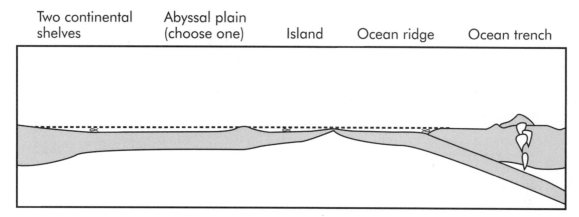

Use the diagram to answer the following questions about features found on the ocean floor.

2. Which landform is created at a convergent plate boundary?

_____ Explain:

3. Which landform is created at a divergent plate boundary?

_____ Explain:

4. Which landforms can form independent of plate boundaries?

_____ _____ _____

Explain:

5. Which landform is the deepest place in the ocean? _____

6. Where is this landform created?   near the center of the ocean   near the edge of the ocean

7. Explain why an island might form in an ocean.

Lecture Tutorials for Introductory Geoscience

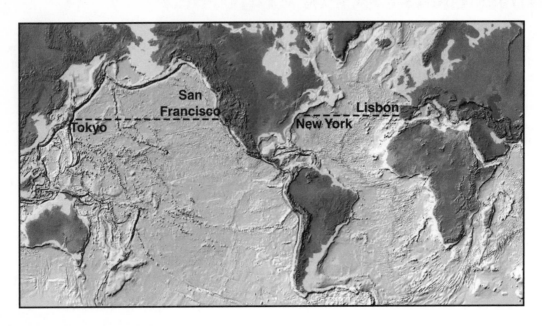

8. If you could walk along the Atlantic Ocean seafloor from New York to Lisbon, Portugal, what landforms would you walk past?

9. If you could walk along the Pacific Ocean seafloor from San Francisco to Tokyo, Japan, what landforms would you walk past?

10. Explain how the landforms on the ocean floor can be used to determine where plate boundaries are located. Give examples to support your answer.

MELTING ROCKS AND PLATE TECTONICS

Part 1: Convergent Boundaries

Below is a cross section showing subduction of an ocean plate beneath a continental plate. The ocean surface is indicated by a dashed line.

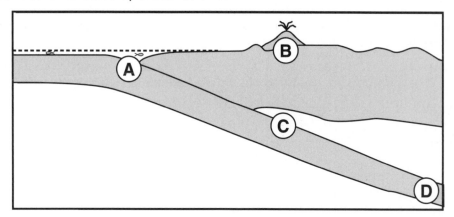

Rock can melt when the temperature increases, the pressure decreases, or water is added. Deeper beneath Earth's surface the temperature increases. As rock rises to Earth's surface, the pressure decreases.

1. Rock cannot melt in the top of the crust because it is not hot enough. Therefore, in what two locations is it hot enough for rock to melt?

   A   B   C   D

2. Volcanoes form directly above where rock melts. Magma is less dense than the surrounding rock, so it rises straight through the crust forming magma chambers for volcanoes. Where could rock melt to form magma that rises into magma chambers and erupts from volcanoes?

   A   B   C   D

3. As the ocean plate subducts and the rocks get hot deep beneath the surface, the minerals in the rocks lose water, adding water to the surrounding rock. Therefore, where could water be added to the rock causing it to melt?

   A   B   C   D

4. Based on your answers to Questions 1, 2, and 3, which location matches all of the requirements? Where do rocks melt to produce magma for the volcanoes?

   A   B   C   D   (circle your answer on the diagram above)

5. Draw an arrow on the diagram to show how the magma moves straight up from its melting location to erupt out of the volcano.

6. Four students are discussing where rock is melting.

    Student 1: *I picked A because that's where the ocean plate is being subducted beneath the continental plate.*

    Student 2: *I picked B because that's where the volcanoes are, so that's where the magma is.*

    Student 3: *I picked C because the volcanoes are right above it, and it's deep enough to be hot enough to melt rock.*

    Student 4: *I picked D because that's where the subducting ocean plate disappears in the diagram, and it's deep enough to be hot enough to melt rock.*

    With which student do you agree? Why?

Part 2: Divergent Plate Boundaries

Below is a cross section of a divergent plate boundary. The ocean surface is indicated by a dashed line.

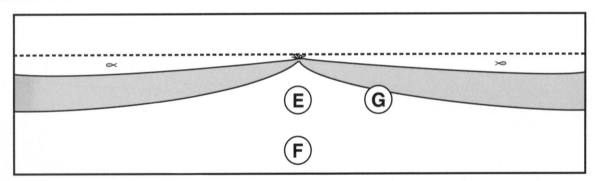

7. Why does rock melt at E but not at F?   hotter temperature   lower pressure   more water

8. Why does rock melt at E but not at G?   hotter temperature   lower pressure   more water

9. Circle the location of melting on the diagram above, and briefly summarize why rock melts in that location.

Part 3: Hotspots

Below is a cross section of a hot spot volcano in the middle of ocean lithosphere.

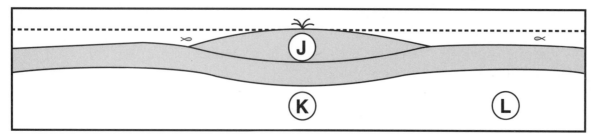

10. Why does rock melt at K but not at J?   hotter temperature   lower pressure   more water

11. Why does rock melt at K but not at L?   hotter temperature   lower pressure   more water

12. Circle the location of melting on the diagram above and briefly summarize why rock melts in that location.

Part 4: Putting It Together

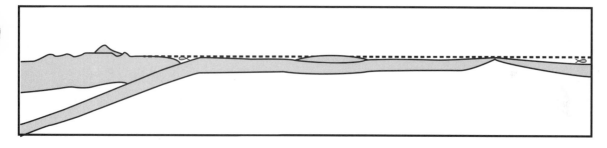

13. On the cross section above, put an "X" exactly at the three locations where rock melts to form magma that will move up and erupt at the surface.

14. Explain why rock melts in these three locations:

    melting location 1:

    melting location 2:

    melting location 3:

15. On the cross section above, draw a star at each of the three places on the surface where lava erupts.

16. Draw an arrow connecting each melting location you labeled with the X with the place that the lava erupts at the surface (labeled with a star).

17. Compare your answers to Questions 13–16 with the work you did earlier in the worksheet; be sure your answers agree.

Lecture Tutorials for Introductory Geoscience

OUTER LAYERS OF EARTH

Part 1: Different Divisions

The diagram below shows the composition and behavior of rocks in the outer portion (couple hundred kilometers or miles) of Earth.

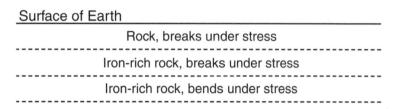

Surface of Earth
_____

Rock, breaks under stress

Iron-rich rock, breaks under stress

Iron-rich rock, bends under stress

----

1. Put a circle ● to the <u>left</u> of the two layers shown above that have the same composition.

2. Put a square ■ to the <u>right</u> of the two layers shown above that have the same behavior under stress.

These are the two common ways of dividing the outer part of Earth into layers. The crust and mantle are layers based on composition, and the lithosphere and asthenosphere are layers based on behavior.

3. Label the two composition layers "crust" and "mantle" on the <u>left</u> side of the diagram above, and bracket the top to the bottom of each layer with this symbol: {

4. Label the two behavior layers "lithosphere" and "asthenosphere" on the <u>right</u> side of the diagram, and bracket the top to the bottom of each layer with this symbol: }

5. A tectonic plate is made up of the lithosphere. Circle the layers of the Earth that are included in tectonic plates.

     crust   upper mantle   lower mantle   lithosphere   asthenosphere

6. What is an important characteristic when determining what makes up a tectonic plate versus what makes up the layer beneath it?

     composition   whether it breaks or bends

7. Is "lithosphere" just a different, more scientific, name for the crust? Explain your answer using the characteristics of each layer.

Part 2: Cross Section

The diagram below represents a slice of the outer parts of Earth. For clarity, the outer layers have exaggerated thicknesses and are not to scale.

8. Fill in the key explaining what the patterns mean with regards to either rock composition or behavior.

9. Label the crust, upper mantle, lithosphere, and asthenosphere. Make sure your labels clearly indicate what is included in each layer.

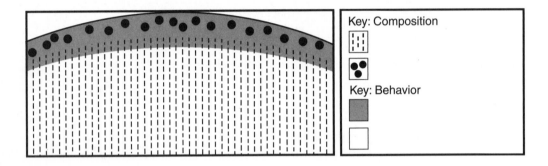

Part 3: Digging to Layers of Earth

10. Two students are discussing the location of the crust compared to Earth's surface.

    Student 1: *I think the crust is right at Earth's surface. We walk around on it and it includes rocks as well as sediments.*

    Student 2: *I think the crust is the solid rock beneath the pieces of rock and sediment beneath the surface. You cannot see the crust.*

    With which student do you agree? Why?

11. If you take a shovel and dig, how deep would you need to go to reach a tectonic plate?

THE MANTLE

The mantle is composed of ultramafic rock, which is made of ferromagnesian minerals (iron- and magnesium-rich, black-colored minerals).

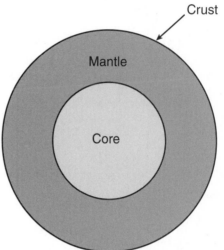

1. If we could drill down to the mantle and pull up a piece of it, what color would the rock be?

   red   black

2. Three students are discussing why many textbooks show the mantle using the color red.

   Student 1: *The mantle is very hot, so textbooks are trying to show that it is so hot by using the color red.*

   Student 2: *Textbooks are showing that the mantle is similar to red-hot lava coming out of volcanoes. It is hot and molten, or liquid.*

   Student 3: *I think that if we were able to get a sample of the rocks from the mantle, it would be red in color, as shown in textbooks.*

   With which student do you agree? Why?

Earthquakes create energy waves which travel through Earth. One type of wave, S waves, travels through solids, but not liquids.

3. Make a prediction: would S waves travel through the mantle if it was liquid?

   S waves would travel through the mantle if it was liquid.

   S waves would not travel through the mantle if it was liquid.

After monitoring earthquakes and the resulting seismic waves, geologists have found that S waves do travel through the mantle.

4. What does this information tell scientists about the phase of the mantle?

   It is solid.   It is liquid.

5. Examine your answer to Question 2. Change it if it does not match your conclusion in Question 4.

Lecture Tutorials for Introductory Geoscience

6. Two students are discussing how much of the mantle is molten.

    Student 1: *The mantle is so hot that a large amount of the rock is melted into liquid. How else would the mantle be able to move during convection? The mantle is half liquid.*

    Student 2: *The mantle moves a couple of inches a year, but it is still nearly all solid, although there may be tiny amounts of melted rock in small cracks. We know this because S waves travel through the mantle.*

    With which student do you agree? Why?

7. To which of the foods listed below is the mantle most similar, based on its physical state (solid versus liquid versus gas) and behavior (flows slowly, flows quickly, does not flow)?

    honey   jello   cracker

    Explain your answer.

THE OUTER CORE

Part 1: Determining the Size of Earth's Outer Core

P and S seismic (earthquake) waves are used to determine the composition and phase of the interior of Earth. The diagram below represents Earth; the star is the location of an earthquake, and the tick marks indicate seismic stations that measure P and S waves.

P waves arrive first and <u>do</u> travel through liquid.

S waves arrive second and <u>do not</u> travel through liquid.

Surface waves arrive last and are the largest.

1. Circle the S waves on each of the seismic stations that recorded S waves. Some have been done for you.

2. Draw arrows representing the path of S waves from the earthquake to each seismic station that recorded S waves. <u>Only draw lines for S waves</u>. Some have been done for you.

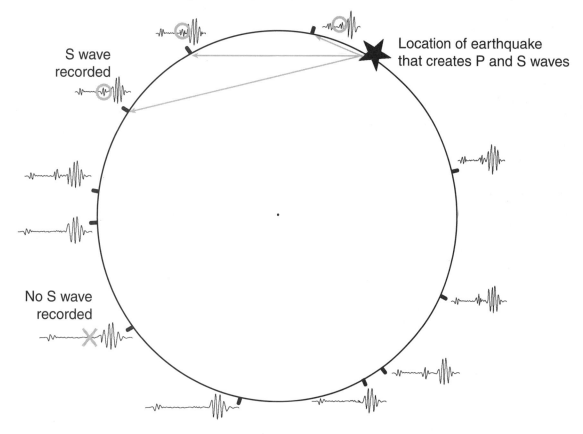

3. Based on the pattern of S waves, what can you determine about the phase of the outer core of Earth? Is it solid or liquid? Explain your answer.

4. Draw the outer core on the diagram, using the distribution of S waves to help you determine the size.

5. The mantle is the layer between the outer core and the crust at the surface of Earth. Based on where S waves are detected around Earth, how much of th e mantle is liquid?

   very little to none   about half   most to all

   Explain your answer.

Part 2: The Outer Core of Other Planets

6. Does this planet have a large or small molten core?

   P = Arrival of P waves
   S = Arrival of S waves

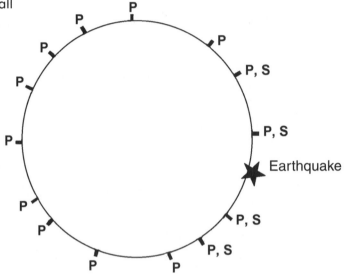

7. What can you conclude about the interior of this planet?

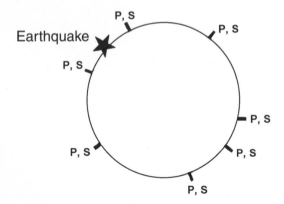

Lecture Tutorials for Introductory Geoscience

MAGMA SOURCE DEPTH

Part 1: Earth's Layers

Earth can be divided into three layers based on composition (what it is made of): the crust, the mantle, and the core. The core can be divided into two layers based on phase: the liquid outer core and the solid inner core. The depths in the chart below are approximate and vary with location.

Layer	Depth of Top	Depth of Bottom	Phase and Composition
Crust	surface of Earth	30 km	solid, rock
Mantle	30 km	2900 km	mostly solid, rock
Outer core	2900 km	5100 km	liquid, metal
Inner core	5100 km	center of Earth	solid, metal

1. Sketch and label the four layers of the Earth on the diagram below. The inner core has been drawn and labeled for you.

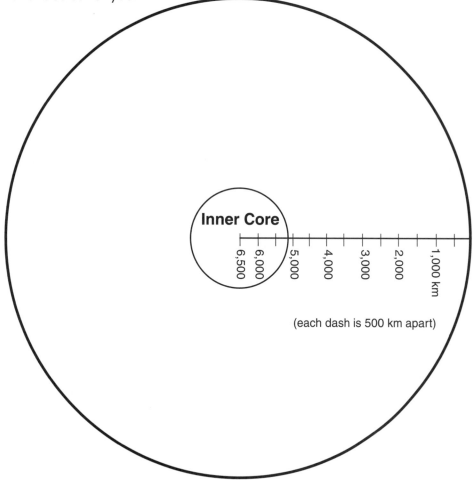

2. What is the best comparison for the thickness of the crust?
   a. The crust has the same relative thickness as the skin of an apple.
   b. The crust has the same relative thickness as the peel of an orange.

Part 2: Origins of Magma

Materials melt with a combination of high temperature and low pressure. If the temperature is too low or the pressure is too high, rocks will not melt.

The outer core has the right combination of temperature and pressure for metal to be molten. At 15 to 100 km below the surface, the temperature and pressure are potentially just right to partially melt rock. This depth is where pockets of most volcanic magma are formed.

3. On the diagram of Earth, draw a star at the depth of the source of magma.

4. What layers melt to form magma?

   crust   upper (outer) mantle   lower (inner) mantle   outer core   inner core

5. Overall, how much of the mantle is liquid?   very little to none   about half   most to all

6. According to your diagram, estimate how far the molten metal from the outer core would have to travel to erupt as a volcano. For comparison, New York and Los Angeles are ~4,000 km apart.

7. What is the composition of the outer core? What is the composition of erupted lava from a volcano?

8. Two students are debating whether the molten outer core erupts as volcanoes.

   Student 1: *I don't think the molten outer core erupts as volcanoes because the magma would have to travel thousands of kilometers through the mantle to reach the surface, and I don't think it could go that far through the mostly solid mantle.*

   Student 2: *If the outer core erupted as volcanoes, then we would have pure metal erupting out of Earth's surface. Volcanoes erupt molten rock, so the molten source cannot be the outer core.*

   Do you agree with one or both students? Why?

9. You are the science advisor to a movie. The screenwriters come to you with the following scenario: A mad scientist threatens to detonate a bomb in the center of Earth, triggering volcanoes to erupt around the world unless world leaders pay him a large ransom. Explain to the screenwriters why their story is or is not scientifically accurate. (Note: Atomic bombs at Earth's surface can cause damage up to 20 km away.)

Lecture Tutorials for Introductory Geoscience

HOT SPOTS

Part 1: Hot Spot Islands and Plate Movement

The diagram below shows several volcanic islands formed by a hot spot and the age of the youngest volcanic rock (Ma = millions of years ago). Use the diagram to answer the questions.

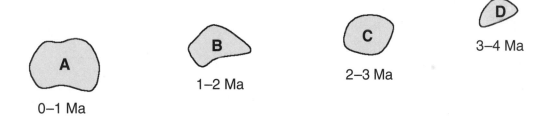

A   0–1 Ma
B   1–2 Ma
C   2–3 Ma
D   3–4 Ma

1. Where is the hot spot now?   A   B   C   D

2. Which of these locations has active volcanoes?   A   B   C   D

3. Which arrow below best shows the direction the plate is moving over the stationary hot spot?

4. Two students are discussing the movement of the plate over the hot spot.

   Student 1: *The hot spot does not move, and the plate moves over it, so I chose Arrow A. If you draw a spot on the table to represent the hot spot, and this paper represents the tectonic plate, then you would have to move the paper (lithosphere) in the direction of Arrow A to get the right ages for the islands.*

   Student 2: *The hot spot is moving from island to island beneath the surface, so I chose Arrow B. If you draw arrows from the oldest island to the youngest island, then the arrows match Arrow B.*

   With which student do you agree? Why?

5. Put a circle around the location where an island will appear in the future, and label it "next island."

Hawaii is an island chain formed by a hot spot. Below is a map of Hawaii, with the approximate location of the hot spot given by a star.

6. Which island is made up of the youngest rock?

   Kauai   Oahu   Maui   The Big Island

7. Which way is the Pacific tectonic plate moving over the hot spot?

   A   B

8. Two students are discussing where they would go on vacation to Hawaii.

   Student 1: *I want to take hikes in tropical areas, and I really want to see lava flowing, so I would vacation on the Big Island.*

   Student 2: *I've heard great things about hikes in Maui, so I think you should vacation on Maui to hike and see lava flowing.*

   With which student do you agree? Why?

9. Your friends move to Kauai. Should they get volcano insurance? Explain why or why not, using information about hot spots and tectonic plate motion.

Part 2: Hot Spot Islands versus Convergent Plate Boundary Islands

The diagram below shows several volcanic islands formed by an ocean-ocean convergent plate boundary and the age of the youngest volcanic rock (Ma = millions of years ago). Use the diagram to answer the questions.

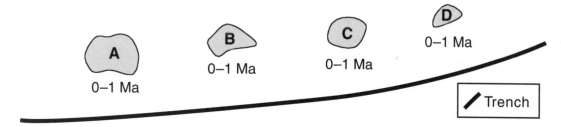

10. Which of these locations has active volcanoes?   A   B   C   D

11. How many tectonic plates are in this diagram? _____

12. Which arrow below best shows the relative direction the plate with the islands is moving?

13. If you saw a chain of volcanic islands in the middle of the ocean, give at least two pieces of information that would be helpful to determine whether they were formed by a hotspot or a convergent plate boundary.

ROCKS AND THEIR FORMATION

MINERAL GROUPS

1. Examine the minerals and their corresponding chemical formulas (see chemical symbols below). Circle the key aspects of the chemical formula and determine which mineral group each mineral falls into (use the groups listed below).

Chemical Symbols		Groups of Minerals ⟶ Description
C	Carbon	• Silicates (has silicon and oxygen)
Fe	Iron	• Non-ferromagnesian silicates (no iron, no magnesium) ⟶ Light color, not metallic
Mg	Magnesium	• Ferromagnesian silicates (also has iron and/or magnesium) ⟶ Dark color, not metallic
O	Oxygen	• Carbonates (has carbon and oxygen: CO_3) ⟶ Often fizzes w/acid
Si	Silicon	• Other (no silicon, often a metal, salt or sulfate) ⟶ Sometimes metallic

Mineral Name, Chemical Formula, and Description	Key Aspects of the Chemical Formula (circle or cross out)			Mineral Group (circle one)
Quartz $Si\,O_2$ Light colored, not metallic	Si Fe	C Mg	O	non-Fe Mg silicate Fe Mg silicate carbonate other
K-feldspar $K\,Al\,Si_3\,O_8$ Light colored, not metallic	Si Fe	C Mg	O	non-Fe Mg silicate Fe Mg silicate carbonate other
Calcite $Ca\,C\,O_3$ Light colored and reacts with acid, not metallic	Si Fe	C Mg	O	non-Fe Mg silicate Fe Mg silicate carbonate other
Biotite $K\,(Mg,Fe)_3\,Al\,Si_3\,O_{10}\,(OH)_2$ Dark colored, not metallic	Si Fe	C Mg	O	non-Fe Mg silicate Fe Mg silicate carbonate other
Pyrite $Fe\,S_2$ Metallic	Si Fe	C Mg	O	non-Fe Mg silicate Fe Mg silicate carbonate other
Halite (Salt) $Na\,Cl$ Light colored, not metallic	Si Fe	C Mg	O	non-Fe Mg silicate Fe Mg silicate carbonate other

2. Two students are debating the classification of graphite (chemical formula: C).

   Student 1: *I think that graphite belongs in the "Other" category. It has carbon, but it doesn't have oxygen, so it's not a carbonate.*

   Student 2: *But the word "carbonate" has the word "carbon" in it! I think that if any mineral has carbon in its chemical formula, it must be a carbonate.*

   With which student do you agree? Why?

3. Pencil lead is made of graphite. What about the appearance of graphite indicates that it is in the "Other" mineral group?

4. What is the classification of hematite (Fe_2O_3)? Explain your answer using the key elements in the chemical formula.

5. Muscovite mica [$KAl_2AlSi_3O_{10}(OH)_2$] has nearly the same chemical formula as biotite mica [$K(Mg,Fe)_3AlSi_3O_{10}(OH)_2$]. They both form crystals that are platy sheets. Based on their chemical formula, predict how they are different in appearance.

6. The igneous rock gabbro is made up primarily of ferromagnesian silicate minerals. Predict what gabbro looks like. Explain your answer.

Lecture Tutorials for Introductory Geoscience

MINERALS AND ROCKS

Below are diagrams showing atoms, minerals, and a rock.

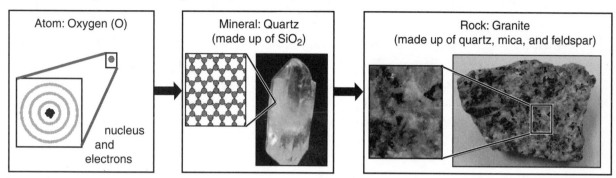

1. What are minerals made up of?   atoms   other minerals   rocks

2. Minerals are the building blocks of what?   atoms   other minerals   rocks

3. Rank minerals, rocks, and atoms in terms of size, from smallest to largest.

   Smallest _____   _____   _____ Largest

4. Minerals have a definite crystal structure, meaning that the atoms are arranged in a particular pattern. Examine the picture of granite above. Does this also apply to rocks?

   Yes   No

Minerals have an exact chemical composition, meaning that there is the same ratio and arrangement of atoms throughout (e.g., 1 silicon for every 2 oxygen).

5. Examine the picture of granite above. Does rock also have an exact composition?

   Yes   No

6. Which best describes minerals?

   the same pattern throughout (homogenous)   varies throughout (heterogeneous)

7. Which best describes rocks?

   the same pattern throughout (homogenous)   varies throughout (heterogeneous)

8. Two students are discussing a piece of quartz they found on the ground.

   Student 1: *I think it is a rock because it is a hard, solid material from the ground.*

   Student 2: *I think it is a mineral because it is the same throughout, and it is shiny.*

   With which student do you agree? Why?

9. In your own words, how would you explain the difference between a rock and a mineral?

ROCK CATEGORIES

Categories are used by people to help simplify information because the characteristics that define categories can be applied to whatever is in that category. The way categories are defined should be useful.

For example, biologists determine relationships between animals. They could categorize animals as "big" and "small," but biologists find that the categories "mammal" and "fish" are more useful when determining relationships between animals.

Four characteristics that define the category mammals are:

- give birth to live young
- have hair
- have lungs
- make milk for their young

1. What are two characteristics that define the category fish?

2. Dolphins are mammals, but your friend says that dolphins are fish. What is one physical feature your friend would incorrectly believe dolphins have?

One way to divide rocks into categories would be to divide them based on density (approximately heavy versus lightweight for the same size rocks).

3. In addition to density, name two different characteristics that can be used to divide rocks into different categories. These do not need to include geologic terminology.

4. A geologist wants to determine if there was a volcano in a particular area by looking at the rocks there. How useful are your listed classifications (in Question 3) for determining the geologic history of that area?

    very useful   not very useful

Geologists have found that the most useful way to divide rocks into categories is to create categories based on how the rocks formed. These categories give geologists valuable information about the environment of formation. These three categories are:

Igneous rocks: Formed when magma cools and solidifies

Sedimentary rocks: Formed when pieces of rock (sediments) are cemented together

Metamorphic rocks: Formed when atoms in a rock change configuration

5. Granite is an igneous rock. Based on that classification, how does granite form?

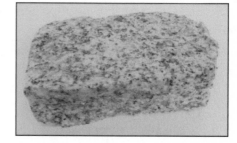

6. Does granite contain sediments? Explain your answer.

7. Three students are discussing whether or not lava initially coming out of a volcano contains abundant sediments.

> **Student 1:** *No, hot lava does not contain sediments because the sediments would melt in the lava. When it cools, the igneous rock is made out of minerals that are cooled and crystallized lava.*

> **Student 2:** *Yes, as it comes out of a volcano, lava picks up sediments, which somehow do not melt. When it cools, the sediments are trapped inside and outside of the igneous rock.*

> **Student 3:** *Yes, magma is made up of other rocks that melted, such as sedimentary rocks. So it can contain sediments.*

With which student do you agree? Why?

8. Gneiss is a metamorphic rock. Explain why gneiss cannot be made up of melted rock.

THE ROCK CYCLE

Part 1: The Rock Types

The three types of rocks are igneous, sedimentary, and metamorphic rocks.

1. Briefly describe how each of the rock types forms.

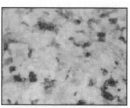 Igneous:

 Sedimentary:

 Metamorphic:

2. Can igneous rocks form from the following rock types?

   igneous   Y or N   sedimentary Y or N   metamorphic Y or N

3. Can sedimentary rocks form from the following rock types?

   igneous   Y or N   sedimentary Y or N   metamorphic Y or N

4. Can metamorphic rocks form from the following rock types?

   igneous   Y or N   sedimentary Y or N   metamorphic Y or N

5. Two students are debating the answers to Questions 2–4.

   Student 1: *I think that the different rock types can form from the other rock types, but they can't form from themselves. For example, igneous rocks can form from sedimentary and metamorphic rocks, but not other igneous rocks.*

   Student 2: *Why not? If metamorphic rocks and sedimentary rocks can melt and form igneous rocks, why can't igneous rocks melt and form igneous rocks again? I think that all rock types can form from all other rock types. So I circled Y for everything.*

   With which student do you agree? Why?

Part 2: The Rock Cycle

6. The diagram below shows the three types of rocks. The arrow showing how sedimentary rocks become metamorphic rocks through increasing heat and pressure is drawn below. Draw the rest of the arrows and label them showing how one type of rock can change into another type of rock. Make sure you have an arrow for each of the changes you thought could happen in Questions 2–5. This is a diagram of the rock cycle.

    Some of the following terms might be useful: melt into magma, cool from magma into rock, erode into sediments, deposit sediments, transform sediments into rock, increase heat, increase pressure, change.

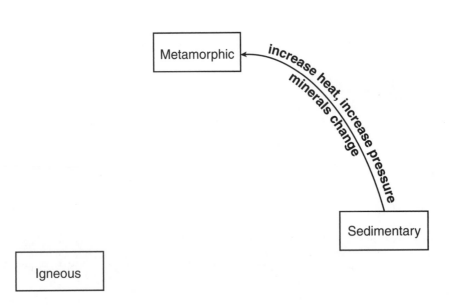

7. Many people describe the rock cycle as rocks going from igneous to sedimentary to metamorphic and then back to igneous again. Explain why this does <u>not</u> fully explain the rock cycle.

ROCK TYPES ON OTHER PLANETS

On Earth there are igneous, sedimentary, and metamorphic rocks. Understanding the way they formed allows us to determine whether we are likely to find them on other planets.

1. Match the rock type with the correct statement describing its formation.

    _____ Formed where the atmosphere or liquid water causes erosion and movement of rock pieces.

    _____ Found mostly near convergent tectonic plate boundaries where the temperature and pressure can be very high.

    _____ Found in places where the interior is so hot that rock melts and then cools again to form new rock.

On the Moon, the first rocks to form when it was molten were the outermost rocks that cooled to form the light-colored highlands. Then, molten rock filled in the lower areas and cooled. This rock is called the mare basalt.

2. Based on these descriptions, of what type of rock are the highlands and mare basalt composed?

    Highlands: (circle one)   igneous   sedimentary   metamorphic

    Mare basalt: (circle one)   igneous   sedimentary   metamorphic

Planet	Water	Atmosphere	Molten Interior	Plate Tectonics
Mercury	no	no	early only	no
Venus	no	thick	yes	no
Earth	liquid, ice	medium	yes	yes
Moon	no	no	early only	no
Mars	ice	thin	yes	no

3. Where in our solar system might we find igneous rocks? Explain your choice based on what factors are necessary for an igneous rock to form.

4. Where in our solar system might we find sedimentary rocks? Explain your choice based on what factors are necessary for a sedimentary rock to form.

5. Where in our solar system might we find metamorphic rocks? Explain your choice based on what factors are necessary for a metamorphic rock to form.

6. What is the most common rock type in the solar system? _____

IDENTIFYING IGNEOUS ROCKS

Granite

Gabbro

Rhyolite

Basalt

Part 1: Similar Rocks

1. Divide the four igneous rocks shown above into two groups of your choosing. Circle the rocks that are grouped together with this method.

2. What <u>characteristic</u> did you use to determine which rocks belong in each group?

3. Using different rock characteristics, divide the rocks up again into two different groups.

4. This time, what <u>characteristic</u> did you use to determine which rocks belong in each group?

5. Compare the characteristics you used with other students' characteristics. If you have different characteristics, convince the other students that the two characteristics that you used to divide the rocks are the best two characteristics.

6. After your discussions, list below two ways to divide these rocks into groups.

Part 2: Color and Mineral Size

7. The four ovals in the diagram below represent the four different types of igneous rocks shown in the previous questions. Where the ovals overlap, fill in the characteristics shared by both of those rock types.

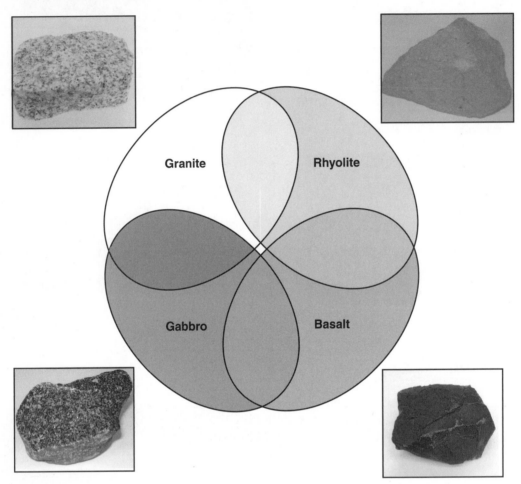

IGNEOUS ROCK MINERAL SIZE

Part 1: Forming Minerals

The size of minerals in an igneous rock is determined by how long the magma takes to cool. To illustrate, everyone should stand up and scatter throughout the room.

1. You have two seconds to form groups as big as possible.   How many per group? _____

2. Scatter again. Now you have 10 seconds.   How many per group? _____

3. Two students are debating about how this activity relates to mineral size in rocks.

    Student 1: *It seems to me that with a longer amount of time, it is possible for all the atoms to form really large minerals.*

    Student 2: *I don't know, I would think that more time means that more minerals will form, and only a little bit of time means only a few big minerals will form.*

    With which student do you agree? Why?

Part 2: Mineral Formation Location

Two bodies of magma are shown in a cross section below. One is above ground and the other is deep within the crust. The length of the arrows represents the rate at which heat is escaping from the molten rock as it cools.

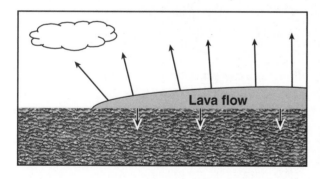 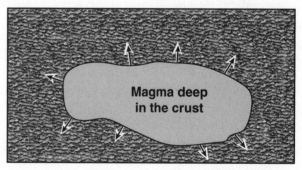

4. Which will cool faster?   lava erupted onto the surface   magma deep underground

5. The igneous rocks granite and gabbro have large minerals. In which location would they have formed?

    on the surface   deep in the crust

6. The igneous rocks rhyolite and basalt have minerals so small it is difficult to distinguish them with the naked eye. In which location would they have formed?

    on the surface   deep in the crust

7. Circle the two rocks that formed deep in the crust.

Granite   Rhyolite   Basalt   Gabbro

Check your answer with your answers for Questions 5 and 6.

Part 3: Porphyry

8. The igneous rock to the right has large, black-and-white colored minerals and many small, gray minerals. You can tell it is an igneous rock because the minerals are rectangular and not rounded like sediments. How might the igneous rock shown to the right have formed?

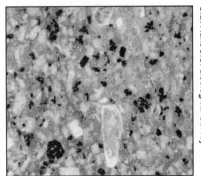

C.B. Hunt/U.S. Geological Survey

Porphyry

9. Two students are debating about the cooling rate of this rock, and the formation of the large minerals.

Student 1: *The magma must have gotten large pieces of sediments that we can see trapped in it, and the sediments didn't melt, even though they were in the magma. So, this rock formed because large pieces of sediment got picked up by lava, and then that lava cooled quickly.*

Student 2: *This is an igneous rock, so everything started off as magma. The large minerals must have formed deep underground when the magma was cooling slowly, like in a magma chamber. But the rest of the rock has very small minerals, so they cooled quickly at the surface.*

With which of these students do you agree? Why?

10. Student 2 said that the large minerals formed deep underground, like in a magma chamber, and the small minerals formed at the surface. Describe what actually happened to form the rock. In other words, what story does the appearance of this rock tell us about its history?

(Hint: In what situation is magma in a magma chamber moved to the surface?)

BOWEN'S REACTION SERIES

To help understand partial melting of rock, imagine a single solid frozen block of dark chocolate chips, ice chips, and butter pieces compressed together. We will compare this frozen block to a hypothetical rock with quartz, potassium feldspar, and amphibole minerals.
This single frozen block is warmed up very slowly.

1. What parts of the frozen block melt first?   chocolate chips   ice chips   butter pieces

2. What melts second?   chocolate chips   ice chips   butter pieces

3. The heating of the block is stopped at a temperature where only the ice is just barely melted but is still near freezing. What happens to the butter pieces and chocolate chips?

    They melt too.   They float around in the melted material.

4. Two students are discussing what happens to the chocolate chips.

    Student 1: *The chocolate chips will melt along with the ice. If they're surrounded by melted material, they should melt as well, so everything will melt together.*

    Student 2: *No, they melt at a higher temperature, and the ice does not reach a high enough temperature when it melts. The melted ice is still much too cold to melt the chocolate chips, so the chocolate chips will sit in whatever is melted.*

    With which student do you agree? Why?

5. In the box below, fill in water ice, butter, and chocolate chips, showing the relative temperatures at which they melt, from lowest temperature to highest temperature.

    (Cool ≈ 0°C)   (Hot ≈ 34°C)

Bowen's reaction series describes the order in which minerals in magma melt and solidify. The order in which the minerals melt (or solidify) is given below from lowest temperatures to highest temperatures.

(Cool ≈ 600°C)   (Hot ≈ 1200°C)

Quartz   Muscovite Mica   Potassium Feldspar   Biotite   Amphibole   Pyroxene   Olivine

A hypothetical igneous rock with quartz, potassium feldspar, and amphibole is heated up.

6. In the rock, which mineral melts first (at the lowest temperature like the ice chips)?

   quartz   potassium feldspar   amphibole

7. Which mineral melts last (at the highest temperature like the chocolate chips)?

   quartz   potassium feldspar   amphibole

8. Give an example of a mineral that could float in magma without melting.
   Explain your choice.

The reverse of Bowen's reaction series happens when magma cools (or freezes) to form minerals.

9. At what temperature does water ice melt? _____

10. At what temperature does liquid water crystallize or freeze? _____

11. What is the relationship between melting temperature and freezing temperatures?

    Melting temperature is higher.   Freezing temperature is higher.   They are the same.

12. Based on your answer to the previous question, the mineral with the lowest melting temperature will be the same as the mineral with the...

    highest crystallization   lowest crystallization
    (freezing) temperature.   (freezing) temperature.

13. For the solid frozen block of chocolate chips, ice chips, and butter pieces compressed together, which has both the lowest freezing temperature and lowest melting temperature?

    chocolate chips   ice chips   butter pieces

14. On Bowen's reaction series shown below, label the following temperatures by drawing four lines connecting the temperature to the appropriate minerals.

    Lowest melting   Highest melting   Lowest crystallization   Highest crystallization
    temperature   temperature   (freezing) temperature   (freezing) temperature

    Quartz   Muscovite Mica   Potassium Feldspar   Biotite   Amphibole   Pyroxene   Olivine

The diagram below shows a microscopic view of minerals in an igneous rock. The minerals that form first as magma cools are nicely shaped and do not run into other minerals because they formed when everything else was still magma. The minerals that form last are irregularly shaped because, as magma cools, they fill in the gaps left behind by the minerals that formed first.

15. Match the mineral name with the mineral.

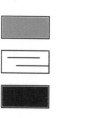   amphibole

          muscovite mica

          quartz

VOLCANO TYPES

Different types of lava erupt to form different types of volcanoes. The type of volcano can indicate if the eruptions will be explosive or peaceful.

Felsic lava has a lot of silicon and oxygen chains and very little iron and magnesium. (Imagine tangled spaghetti strands.)

Mafic lava has a lot of iron and magnesium that break up the silicon and oxygen chains. (Imagine elbow macaroni.)

1. Which type of lava would be runny?   mafic   felsic

2. Which type of lava would be viscous (thick)?   mafic   felsic

Below are two simplified profiles of volcanoes. The profile of a volcano can give us a clue if it is built from thick, viscous lava or runny lava (think about toothpaste versus honey).

3. On the diagram below, write "runny" next to the volcano that is formed from runny lava and "viscous" next to the volcano that is formed from viscous lava.

4. On the diagram below, write "mafic" next to the volcano that is formed from mafic lava and "felsic" next to the volcano that is formed from felsic lava.

The igneous rock rhyolite is formed from felsic lava, which is why it is light in color. The igneous rock basalt is formed from mafic lava, which is why it is dark in color.

5. On the diagram below, write "rhyolite" next to the volcano composed of rhyolite and "basalt" next to the volcano composed of basalt.

Volcanoes will erupt explosively if the gasses in the lava cannot easily escape. Runnier lava will allow the gas to bubble out peacefully, so there is no explosion. However, viscous lava will not allow the gas to escape, so the pressure builds until an explosion releases the gas.

6. Which type of lava will easily let gasses escape?   felsic lava   mafic lava

7. Write "explosive" next to the volcano that will erupt explosively and "peaceful" next to the volcano that will erupt peacefully.

shield volcano

composite volcano   Not to scale

8. Based on the type of eruption, which volcano would you rather live next to—a shield volcano or composite volcano? Explain your answer.

9. Two students are debating which volcano they would rather live next to.

   Student 1: *I would prefer to live next to a composite volcano because the lava is so thick and viscous that they don't have long lava flows. The lava flows will slow down and stop before they get to my house.*

   Student 2: *But, composite volcanoes have explosions that can't be predicted, so they are more dangerous than runny lava. I would prefer to live next to a shield volcano, where the runny lava flows in predictable patterns.*

   With which of these students do you agree? Why?

10. The photograph below is of a volcano in the United States. What can you determine about the volcano based on the picture (e.g., type of volcano, lava, rock, eruption...)?

Hoblitt/U.S. Geological Survey

VOLCANOES ON OTHER PLANETS

Part 1: Observing Volcano Distribution

When a scientist makes a discovery, it helps to have as many different sources of information as possible confirm that discovery. Here we will look at two ways to determine the types of volcanoes on other planets.

1. Examine the maps of volcanoes on Mars, Venus, and Earth. Take one minute to determine if there is a clear pattern in the location of volcanoes on each planet, or if they are distributed in random groups. If there is a pattern, describe what kind of pattern you see.

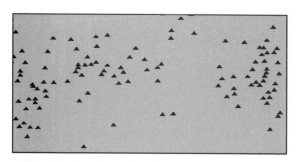

Volcanoes on Venus (triangles)

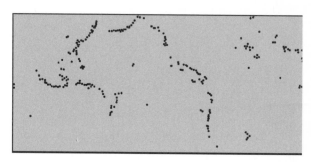

Volcanoes on Earth (dots)

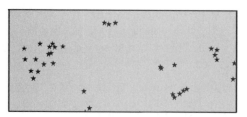

Volcanoes on Mars (stars)

Part 2: Analysis

Volcanoes form either at randomly distributed hotspots or lined up along tectonic plate boundaries. A single planet might have both types of volcanoes.

2. Why do the volcanoes on Earth form where they do?   hot spots   plate tectonics

   Explain how your answer is related to your observations about the maps.

3. Why do the volcanoes on Venus form where they do?   hot spots   plate tectonics
   Explain how your answer is related to your observations about the maps.

4. Why did the volcanoes on Mars form where they did?   hot spots   plate tectonics
   Explain how your answer is related to your observations about the maps.

5. Which planet(s) has/have plate tectonics?   Venus   Earth   Mars

Part 3: Comparing Individual Volcanoes

Another way to determine the cause of volcanoes on other planets is to compare the two types of volcanoes on Earth with volcanoes on other planets. Composite volcanoes (e.g., Mount St. Helens) usually form at plate tectonic boundaries and have steep slopes; shield volcanoes (e.g., Hawaii) usually form at hot spots and have gentle slopes.

6. Look at the profile of volcanoes on Earth drawn to scale. Label each volcano as "composite volcano" or "shield volcano" and indicate if the volcano formed at a hot spot or plate tectonic boundary.

120 km   20 km

Below is a satellite image of Olympus Mons, an example of a volcano on Mars. This volcano is approximately 25 km tall and 600 km wide. It is possible to use satellite information to create a profile like those of Earth shown above.

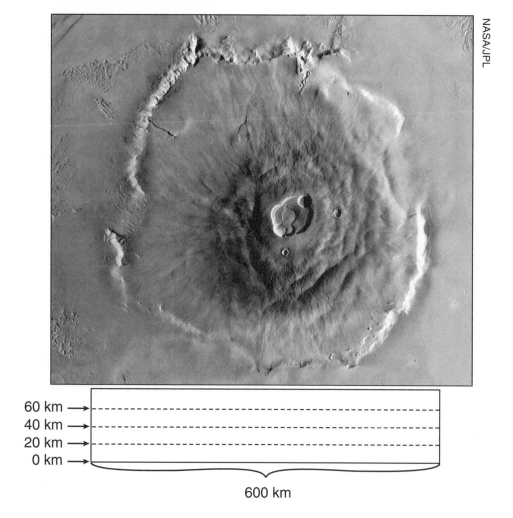

NASA/JPL

60 km →
40 km →
20 km →
0 km →

600 km

7. Use the information about the height and width of the volcano to sketch the profile of this volcano (like those of volcanoes on Earth shown in Question 6) on the graph above.

8. Based on the profiles, is the volcano on Mars a composite volcano or a shield volcano? Explain.

9. Based on the profiles, why did the volcanoes on Mars form?   hot spots   plate tectonics

   Explain how your answer is related to the profile and type of volcano.

You used two methods to determine the type and origin of volcanoes found on Mars: the distribution of volcanoes to determine the likely source of volcanism, and the profile of an individual volcano.

10. Do your two data sets agree? If they do not agree, what might cause the difference?

11. Why is it helpful for a scientist to have two or more different data sets when giving evidence to support a discovery?

WEATHERING

Part 1: Chemical and Physical (Mechanical) Weathering

1. You put salt (the mineral halite) in water. After 10 minutes can you see the salt in the water?

   Yes   No   Explain what happens to the salt.

2. You put sand (the mineral quartz) in water. After 10 minutes can you see the sand in the water?

   Yes   No   Explain how the sand in the water is different than the salt.

Examine the two diagrams below of salt and quartz minerals and what happens to them when they are weathered in water.

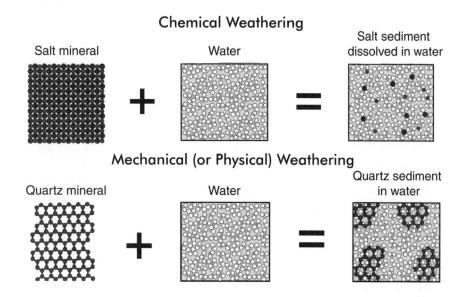

3. For each statement, put an X in the correct column for the type of weathering it best describes.

	Chemical Weathering	Mechanical (or Physical) Weathering
Individual atoms in water		
Little pieces of rock in water		
Results in a change in the chemistry of the mineral		
Resulting sediments are visible in water		
Resulting sediments are not visible in water		

As a result of weathering, sediments form. Therefore, sediments can be a result of chemical weathering or mechanical weathering.

4. Ocean water in coral reefs is clear, but the water still contains sediments. Are these sediments a result of chemical weathering or physical (mechanical) weathering? Explain your answer.

5. If you see a muddy river full of mud and sand, does that river contain sediments that are a result of primarily chemical weathering or physical (mechanical) weathering? Explain your answer.

Part 2: Parent Material

The parent material is one important factor in determining the rate of chemical weathering.

6. For each statement, put an X in the column for the type of weathering each mineral tends to experience. If a mineral does not experience one type of weathering, it will experience the other.

	Chemical Weathering	Mechanical (or Physical) Weathering
Calcite dissolves in water.		
Ferromagnesian silicate minerals (such as biotite, mica, or olivine) rust in water.		
Non-ferromagnesian silicate minerals (such as quartz) are chemically stable in water.		

7. A beach has a supply of sand grains composed of calcite, ferromagnesian silicate minerals, and non-ferromagnesian silicate minerals. It if undergoes lots of chemical weathering, which sand grains will be quickly chemically weathered away?

    calcite   ferromagnesian silicate minerals   non-ferromagnesian silicate minerals

8. After a long time passes, which sand grains will be left on the beach and not chemically weathered away?

    calcite   ferromagnesian silicate minerals   non-ferromagnesian silicate minerals

9. Explain to a friend why many beaches commonly have light, tan-colored sand.

WEATHERING RATES AND SOIL FORMATION

Part 1: Rate of Weathering

Rocks weather at different rates and the amount of weathering depends on several variables:

- <u>The type of rock</u> (e.g., gneiss and granite are generally more resistant; mudstone and basalt are less resistant.)

- <u>The climate</u> (more humid climate results in faster weathering)

- <u>The time of exposure</u> (longer time results in more weathering)

In each of the questions below, use the information above to determine which situation would have the most weathering. In one of the four questions, the amount of weathering in each situation is the same.

1.   A desert environment where there is little rain.

   or A grassland environment where there is more rain.

   or They are the same.

   Explain your answer.

2.   Granite bedrock exposed on the surface for hundreds of years.

   or Granite bedrock exposed on the surface for thousands of years.

   or They are the same.

   Explain your answer.

3.   A mudstone near the beach.

   or A gneiss near the beach.

   or They are the same.

   Explain your answer.

4.   A gneiss in an area where there are many earthquakes.

   or A gneiss in an area where there are no earthquakes.

   or They are the same.

   Explain your answer.

5. Based on the situation where the amount of weathering was the same, circle the factor below that has the smallest effect on weathering.

   climate   time   rock type   earthquakes

Part 2: Soil Formation

The formation of soil requires weathering of the bedrock and the accumulation of the material that was weathered.

6. For each row, circle which situation would have more soil.

   • granite bedrock   or   mudstone bedrock

   • land uncovered by glaciers 100 years ago   or   land uncovered 10,000 years ago

   • jungle   or   desert

   • steep mountain slope   or   flat valley

7. Rank these three situations as having the most soil, an intermediate amount of soil, or the least soil.

   _____ on Kauai, the oldest Hawaiian Island, made out of basalt

   _____ on 30-year-old rhyolite lava flows on Mt. St Helens

   _____ on 30-year-old basalt lava flows on Kilauea, Hawaii

8. Circle the three factors that helped you rank the amount of soil in the previous two questions.

   time   type of rock at surface   earthquakes   climate

9. Explain why the one factor you did not circle does not influence the amount of weathering and soil formed.

SEDIMENTS AND SEDIMENTARY ROCKS

● ## Part 1: Creating Sedimentary Rocks

1. From the choices, list the steps necessary for a parent rock to become a detrital or chemical sedimentary rock. Each step will be used once.

DETRITAL (e.g., shale/mudstone)

- transportation (water, wind, ice)
- ~~parent rock is broken into smaller pieces~~
- deposition of detrital sediments
- compaction and cementation into rock

CHEMICAL (e.g., limestone)

- transportation (dissolved in water)
- ~~parent rock is dissolved~~
- precipitation as rock
- precipitation as shells
- shells are deposited, compacted, and cemented

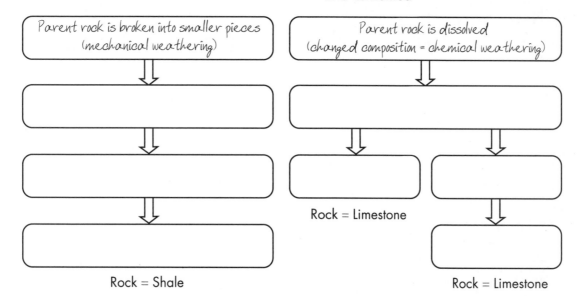

Parent rock is broken into smaller pieces (mechanical weathering)

Parent rock is dissolved (changed composition = chemical weathering)

Rock = Limestone

Rock = Shale

Rock = Limestone

Part 2: Changing from Sediments to Sedimentary Rocks

2. Sandstone is a detrital sedimentary rock made up of sand-sized pieces. Read the four steps below and draw arrows to where they are occurring in the diagram.

- Parent rock is broken into sand-sized pieces (sediments).

- Sand-sized sediments are transported.

- Sand-sized sediments are deposited and buried.

- Deposited and buried sand is compacted and cemented into sandstone (no longer sediments).

Part 3: Sediments versus Sedimentary Rocks

3. One of these photos is a photo of pebble-sized sediments, the other is of a single sedimentary rock. Label the photos. There is a finger for scale in the left photo.

 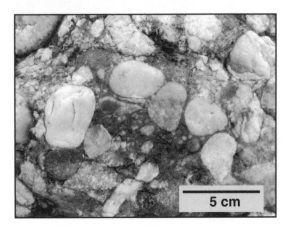

4. What needs to happen so that pebble-sized sediments become a sedimentary rock?

5. Two students are debating about the pebbles in the image on the left and what type of rock the pebbles are.

    Student 1: *In the photo on the left, it is clear that the rocks there are all pebble sized. I think that each pebble is a sedimentary rock because the pebbles have been broken from much larger rocks and they were transported to this new area.*

    Student 2: *I don't agree. Each pebble can be any type of rock—igneous, metamorphic, or sedimentary. Because they have not yet been compacted and cemented together, they have not yet been turned into a sedimentary rock like in the photo on the right.*

    With which student do you agree? Why?

6. You are at the beach with a friend and find pebbles and some slightly larger rocks. Explain why you would not call these sedimentary rocks.

Lecture Tutorials for Introductory Geoscience

SEDIMENTARY DEPOSITIONAL ENVIRONMENTS

Part 1: Rock Types and Features in Different Environments

1. Match the picture of the sedimentary rock to the name of the rock and to the best answer for the environment of deposition. Do this by connecting them with a line.

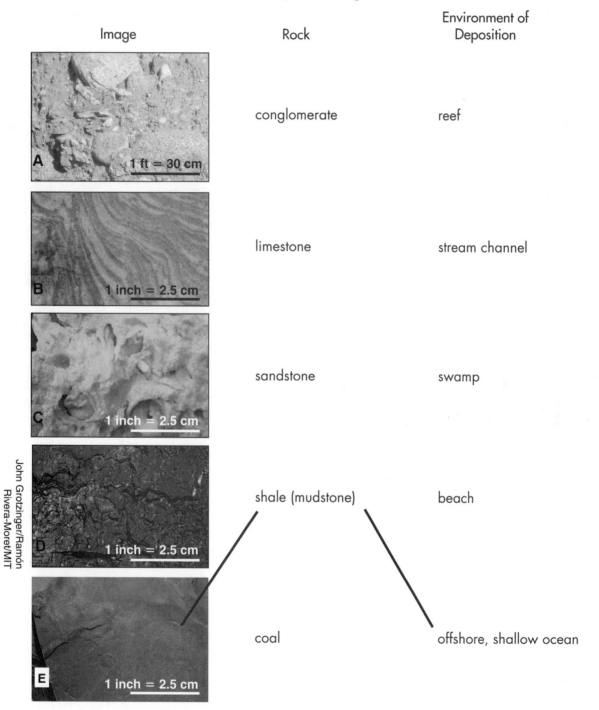

Image	Rock	Environment of Deposition
A (1 ft = 30 cm)	conglomerate	reef
B (1 inch = 2.5 cm)	limestone	stream channel
C (1 inch = 2.5 cm)	sandstone	swamp
D (1 inch = 2.5 cm)	shale (mudstone)	beach
E (1 inch = 2.5 cm)	coal	offshore, shallow ocean

John Grotzinger/Ramón Rivera-Moret/MIT

Lecture Tutorials for Introductory Geoscience

2. Coal mines are common in Pennsylvania. What was the environment of Pennsylvania in the past compared to today?

3. Circle the environment that best matches what Pennsylvania was like in the past.

C.W. Cross/U.S. Geological Survey

4. There is a thick layer of limestone with many shells under Chicago. What can you figure out about the geologic past of that area?

SEDIMENTARY FACIES

Depositional environments vary from location to location. A particular place on Earth's surface can also change depositional environments over time.

1. An area has a layer of sandstone that is interpreted to have once been a beach. On top of that sandstone is a layer of limestone that formed deep in the ocean. What can you say about the sea level of this area?

Limestone
Sandstone

2. Two students are debating about how layers of sandstone and shale (mudstone) relate to rising sea level.

    Student 1: *Sandstone forms from sand deposited right at the ocean's edge and shale forms from sediments deposited in deeper water. So, if the water was shallow first, and then it got deeper, sandstone would be the first layer deposited so it is on the bottom, and shale would be on top.*

    Student 2: *But if the water is getting deeper, that means the shale should be deeper because it forms from sediments deposited in deeper water. So, I think after the sand is deposited the mud that will turn into shale would sink under the sand and form a layer there because it forms in deeper water.*

    With which student do you agree? Why?

3. The diagrams on this page show layers of rock. Which rock layer formed first?

    bottom   middle   top

4. If rock formed from sediments deposited on the land, what rock could make up the layer?

    conglomerate   limestone   sandstone   shale (mudstone)   coal

5. If rock formed from sediments deposited in deep water, what rock could make up the layer?

    conglomerate   limestone   sandstone   shale (mudstone)   coal

6. In the diagram below, write a series of three rock types that could form in an area where the sea level is rising over time. The area started on land and was covered by deeper and deeper water over time. Explain your sequence.

top rock layer:	
middle rock layer:	
bottom rock layer:	

Lecture Tutorials for Introductory Geoscience

METAMORPHIC ROCKS

Part 1: Metamorphic Changes

When rocks are compressed and heated within the Earth they change, or metamorphose. During metamorphism, rocks tend to change in the following ways:

Minerals change into different, more stable minerals.

Minerals grow larger.

Flat minerals align themselves parallel to each other.

1. Examine the photos below. Describe one or two observations how the minerals in these rocks changed during metamorphism.

metamorphism →

Part 2: Metamorphic Grades

During metamorphism, the temperature of and pressure on a rock determine what the rock looks like.

Low-grade metamorphism: If the temperatures and pressures are not very different from those on the surface of Earth, the minerals do not change much and are small.

High-grade metamorphism: High temperatures and pressures cause minerals in a rock to change a lot; minerals grow large, and may form into dark and light bands.

Slate

Schist

Gneiss

2. List the rocks above in order from low-grade to high-grade metamorphic rocks.

Low grade _____ _____ _____ High grade

Lecture Tutorials for Introductory Geoscience

3. Describe what slate would look like if it was metamorphosed more.

4. Three students are thinking about the changes in slate during metamorphism.

   Student 1: *I think it would change to a higher metamorphic grade, so instead of being a low-grade metamorphic rock, it would change to an intermediate-grade metamorphic rock. So it will become schist, and it would have larger minerals.*

   Student 2: *It would look pretty much the same, but it would get even flatter because of the higher pressures during metamorphism. For example, if it started out two inches thick, it might be one inch thick after metamorphism.*

   Student 3: *It would melt, and then new minerals would form.*

   With which student do you agree? Why?

Part 3: Forming Metamorphic Minerals

5. Two students are thinking about how metamorphic minerals form.

   Student 1: *In metamorphic rocks, the minerals melt a little bit due to the extremely high temperatures. The melting causes the atoms to flow around and grow bigger minerals.*

   Student 2: *New minerals form and grow bigger because the rock is getting compressed and heated. The atoms that formed minerals in the parent rock rearrange to form bigger, new minerals.*

   With which student do you agree? Why?

6. Schist contains a lot of the mineral mica, which is what makes it shiny. Slate contains clay, but not much mica. Where do the mica minerals come from as slate metamorphoses into schist?

7. If a rock melts, can it be considered a metamorphic rock? Explain.

8. Use your answer to Question 7 to change your answers to Questions 4 and 5, if necessary.

THE HISTORY OF METAMORPHIC ROCKS

Part 1: Parent Rocks

The parent rock of a metamorphic rock is the original rock before it was changed by metamorphism. It plays an important role in determining the type of resulting metamorphic rock.

1. For each pair below, identify the parent rock or the resulting metamorphic rock. The rocks we will be considering are limestone, shale (mudstone), marble, slate, and gneiss.

Parent Rock   Resulting Metamorphic Rock

limestone

slate

shale (mudstone)

John Grotzinger/Ramón
Rivera-Moret/Harvard
Mineralogical

2. A large area undergoes the same amount of metamorphism (all rocks reach the same metamorphic grade). However, after the metamorphism, some rocks are marble and some rocks are slate. Why?

3. There is a large area made up of schist. What is the most likely environment that existed before the area was metamorphosed?

    desert dunes   coral reef   ocean floor   volcano

4. A metamorphic rock is a "changed" rock. How is it possible to use metamorphic rocks to figure out the geologic history of an area before the area was metamorphosed? Give at least one example in your answer.

Lecture Tutorials for Introductory Geoscience

Part 2: Metamorphism and Plate Boundaries

Two common types of metamorphism are:

<u>Contact metamorphism</u>: high temperatures from being near hot magma

<u>Regional metamorphism</u>: high temperatures and pressures associated with mountain building

For each of the tectonic locations below, circle the type(s) of metamorphism that might occur. Use the space to write a brief explanation.

5. Divergent boundary: contact   regional   none

6. Convergent boundary (ocean-continent):   contact   regional   none

7. Convergent boundary (continent-continent):   contact   regional   none

8. Transform boundary: contact   regional   none

9. The metamorphic rock gneiss forms primarily through regional metamorphism. At what type of ancient plate boundary(ies) would you expect to find gneiss?

10. Two students are discussing what they could determine about a region that is composed of a large amount of marble.

    Student 1: *We could say that the area was once a shallow sea because marble forms from limestone, and limestone forms in shallow seas.*

    Student 2: *We could say that the area was once a convergent plate boundary because the rocks are metamorphosed. If there was no convergent boundary, there is no way that all the rocks would have been metamorphosed.*

    Do you agree with one or both students? Why?

11. In New England, you can find marble, slate, and gneiss. What can you interpret about the geologic history of the area?

METAMORPHIC ROCK FACIES

A metamorphic facies is a group of metamorphic rocks that formed under particular temperature and pressure conditions.

Part 1: Metamorphic Conditions

The diagram to the right shows nine rocks that form at conditions with different pressure and temperature combinations.

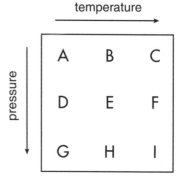

1. Use a diagonal arrow to show how the conditions change as depth increases on the diagram.

2. Circle the conditions under which Rocks B and C would form. You may circle more than one.

   Pressure: low   medium   high

   Temperature: low   medium   high

3. Circle the conditions under which Rocks D and G would form. You may circle more than one.

   Pressure: low   medium   high

   Temperature: low   medium   high

4. Circle the conditions under which Rocks E and I would form. You may circle more than one.

   Pressure: low   medium   high

   Temperature: low   medium   high

5. If rock is metamorphosed by being in contact with magma, what will change more, the pressure on the rock, the temperature of the rock, or both? Explain your answer.

6. Therefore, if rock is metamorphosed by contact with magma, which of the following letters best represents the conditions under which the rock formed, or the metamorphic facies?

   B and C (hornfels)   D and G (blueschist)   E and I (greenschist)

7. If rock is slowly buried very deep within mountain zones, what will change, the pressure on the rock, the temperature of the rock, or both? Explain your answer.

8. Therefore, if rock is slowly buried deep within mountain zones, which of the following letters best represents the conditions under which the rock formed, or the metamorphic facies?

    B and C (hornfels)   D and G (blueschist)   E and I (greenschist)

9. If rock rapidly moves deeper beneath the surface, what will change faster, the pressure on the rock or the temperature of the rock? Explain your answer.

10. Therefore, if rock rapidly moves deeper beneath the surface, which of the following letters best represents the conditions under which the rock formed, or the metamorphic facies?

    B and C (hornfels)   D and G (blueschist)   E and I (greenschist)

Part 2: Plate Tectonics

11. At what two types of plate boundaries does magma move into cooler crustal rocks causing Rocks B and C (hornfels) to form? Explain.

    divergent   convergent   transform

    because...

12. At what type of plate boundary do large mountain zones form causing Rock E and I (greenschist) to form? Explain.

    divergent   convergent   transform

    because...

13. At what type of plate boundary does rock rapidly move deep beneath the surface causing Rocks D and G (blueschist) to form? Explain.

    divergent   convergent   transform

    because...

Lecture Tutorials for Introductory Geoscience

Hornfels (Rocks B and C) and greenschist (Rocks E and I) metamorphic rocks that are hundreds of millions of years old are found in Maine.

14. Based on the metamorphic rocks present, what can you determine about the geologic history of Maine?

Blueschist (Rocks D and G) and greenschist (Rocks E and I) metamorphic rocks that are millions of years old are found near San Francisco, California.

15. Based on the metamorphic rocks present, what can you determine about the geologic history of San Francisco?

THE ROCK CYCLE AND PLATE TECTONICS

The diagram below shows a cross section of the Earth's tectonic plates. The ocean surface is indicated by a dashed line.

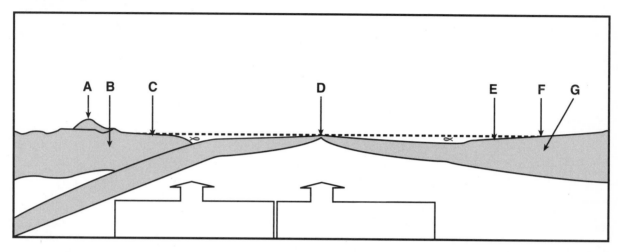

1. In the boxes near the bottom of the diagram, label the two plate boundaries "Convergent with subduction," and "Divergent."

2. Put arrows on either side of each plate boundary showing the direction each plate is moving.

3. On the letters on the cross section above, label a place where each of the rocks listed below forms (some letters will be used more than once and others will not be used):

   _____   _____   _____

   basalt   shale   slate and schist

   granite and diorite   limestone

   andesite and rhyolite   sandstone

4. On the line above each column of rocks in Question 3, label the rock type as igneous, metamorphic, or sedimentary.

5. Two students are discussing where metamorphic rocks are formed.

   Student 1: *The convergent boundary is where the ocean plate is moving toward the continental plate, so the metamorphic rocks form at B.*

   Student 2: *I think they form at F because they both appear to be made out of sand-type sediments, and F is located on a beach.*

   With which student do you agree? Why?

6. In which location(s) might each rock type typically form in terms of tectonic plate boundaries?

Igneous:	divergent	convergent	transform	no plate boundary
Metamorphic:	divergent	convergent	transform	no plate boundary
Sedimentary:	divergent	convergent	transform	no plate boundary

Lecture Tutorials for Introductory Geoscience

GEOLOGIC LANDFORMS AND PROCESSES

TOPOGRAPHIC PROFILES

Below is a topographic map of a small island you will need to answer the questions below. The outer topographic line is at sea level.

Topographic Map

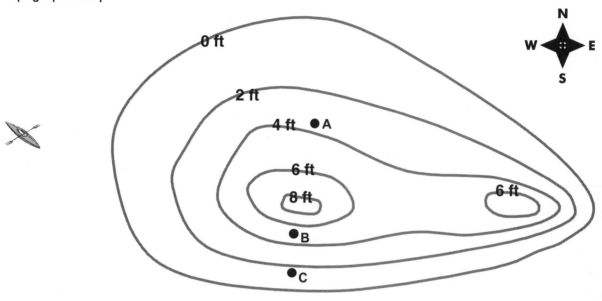

Part 1: Topographic Maps and Steepness

1. Which side of the island is steeper, the west side or the east side? Explain your answer.

2. Two students are debating which side of the island is steeper.

    Student 1: *I think the west side is steeper because it goes from the highest elevation down to sea level. On the east side, the elevation change is smaller because that hill is only 6 feet tall. The big change in elevation on the west side means it's steeper.*

    Student 2: *No, I disagree. It is the east side that is steeper because the steepness is shown by how close contours are together. If they're closer, that means the elevation is changing faster, which means it's steeper.*

    With which student do you agree? Why?

Part 2: Profiles

3. You walk straight from B to C. Would you walk uphill, downhill, or a combination of both?

4. You walk straight from B to A. Would you walk uphill, downhill, or a combination of both?

5. You are in a canoe far to the <u>west</u> of the island (see map). You look back to the east at the profile of the island. Which sketch below best shows what the island would look like? (Circle it.) Explain why the other two choices are not correct.

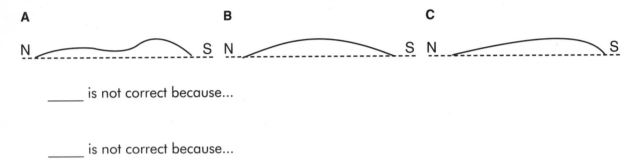

A   B   C

_____ is not correct because...

_____ is not correct because...

Below is an east-west profile across the island (from the south looking to the north). The ocean surface is indicated by a dashed line. Use this profile to answer the following questions.

Profile

6. Is it possible to show an arrow indicating "up" on the profile? If it is possible, then draw in the arrow. If it is not possible, explain why not.

7. Is it possible to show an arrow indicating "north" on the profile? If it is possible, then draw in the arrow. If it is not possible, explain why not.

Lecture Tutorials for Introductory Geoscience

8. Two students are debating what arrows can be drawn on profiles.

    Student 1: *I think that both the arrows can be drawn on the profile because "up" and "north" on maps are similar. You could draw them both pointing in the same direction toward the top of the paper.*

    Student 2: *"Up" is the direction away from ground, and it would point toward the top of the paper, like you said. But "north" and "up" are not the same direction because "north" on the profile points into the paper, so it cannot be drawn.*

    With which student do you agree? Why?

Lecture Tutorials for Introductory Geoscience

PLANET SURFACE FEATURES

Part 1: Planet Features

1. What feature requires a planet to have a hot, molten interior?

   dunes   impact craters   stream beds   volcanic lava flows

2. What feature requires a planet to have an atmosphere?

   dunes   impact craters   stream beds   volcanic lava flows

3. What feature requires a planet to have liquid on the surface?

   dunes   impact craters   stream beds   volcanic lava flows

4. What feature does NOT require a planet to have any particular characteristics?

   dunes   impact craters   stream beds   volcanic lava flows

The following images are of three different planets (Mercury, Earth, and Mars). All the images are of the solid, rocky surface and were taken by NASA spacecraft.

5. Examine these images and identify the type of <u>surface feature</u> shown: sand dunes, impact craters, stream beds, lava flows.

6. For each planet, write down what you can determine about the planet based on those images (if it has an atmosphere, a molten interior, or liquid on the surface).

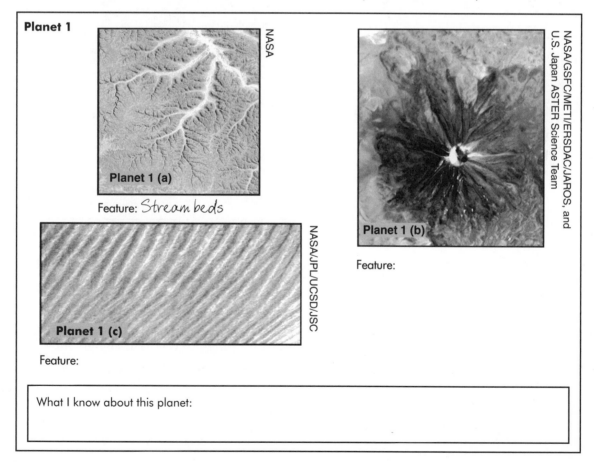

Planet 1

Planet 1 (a)

NASA

Feature: *Stream beds*

Planet 1 (b)

NASA/GSFC/METI/ERSDAC/JAROS, and U.S. Japan ASTER Science Team

Feature:

Planet 1 (c)

NASA/JPL/UCSD/JSC

Feature:

What I know about this planet:

Lecture Tutorials for Introductory Geoscience

Planet Surface Features

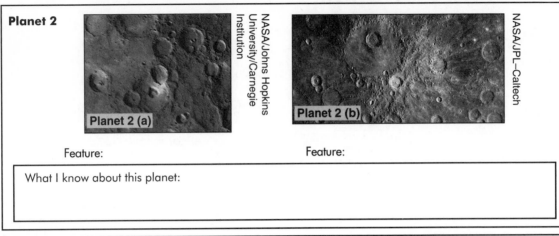

Planet 2

Planet 2 (a)
NASA/Johns Hopkins University/Carnegie Institution

Planet 2 (b)
NASA/JPL–Caltech

Feature: _____   Feature: _____

What I know about this planet:

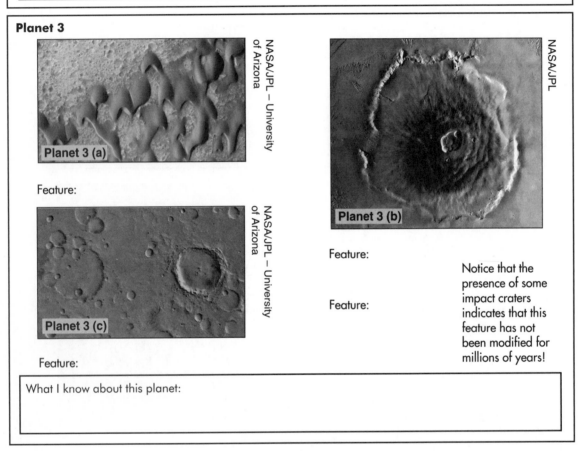

Planet 3

Planet 3 (a)
NASA/JPL – University of Arizona

Feature: _____

Planet 3 (c)
NASA/JPL – University of Arizona

Feature: _____

Planet 3 (b)
NASA/JPL

Feature: _____

Feature: _____

Notice that the presence of some impact craters indicates that this feature has not been modified for millions of years!

What I know about this planet:

Mars has an atmosphere, and it once had a hot, molten interior and liquid water (although not any more). Mercury does not have an atmosphere, liquid water, or a hot, molten interior.

7. Using the information given, label each of the planets as Earth, Mercury, or Mars.

8. The Moon is completely covered in craters. What can you determine about the Moon based on this information?

Lecture Tutorials for Introductory Geoscience

EARTHQUAKE INTENSITY AND MAGNITUDE

Part 1: Magnitude versus Intensity

The <u>magnitude</u> of an earthquake is the amount of energy that is released as the rock breaks. It is the Richter scale number generally displayed by the news.

The <u>intensity</u> of an earthquake is the measure of damage and deaths it caused. A high-intensity earthquake results in a great deal of damage and a high death toll.

1. Describe two situations in which a large-magnitude earthquake can have a low intensity.

2. Describe two situations in which a small-magnitude earthquake can have a high intensity.

3. Two students are debating earthquake intensity and magnitude.

    Student 1: *An example of a large-magnitude earthquake that has a low intensity is if it hit an area with a low population.*

    Student 2: *So, if a really, really large-magnitude earthquake hit the desert in California where nobody lives, you're saying it would not have a high intensity? I don't agree. If the earthquake is that big, I think it would need to have a high intensity as well.*

    With which student do you agree? Why?

Part 2: Earthquake Intensities

The 7.1 magnitude Loma Prieta earthquake (1989) caused 63 fatalities and $10 billion in damage. There was heavy damage near the fault, but there was more damage approximately 50 miles away at the edges of the heavily populated San Francisco Bay. Many buildings and bridges built on soft sediments near the bay were destroyed.

4. The Loma Prieta earthquake is an earthquake with a high intensity. Explain why the earthquake had such a high intensity. There will be more than one reason.

H.G. Wilshire/U.S. Geological Survey

LOCATIONS OF EARTHQUAKES

Part 1: Earthquake Patterns

The USGS map below shows the locations of earthquakes around the world over a 27-year period.

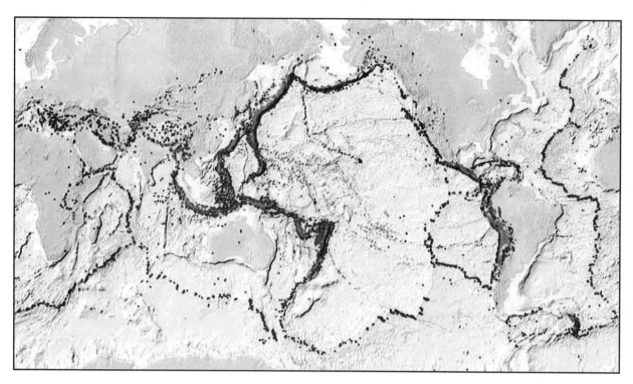

Take a careful look at the pattern of earthquakes scattered around Earth's surface.

1. Do earthquakes occur around the coastlines of <u>all</u> oceans?   Yes   No

    If you chose "No," give an example of an ocean with no earthquakes around the edges and mark it on the map.

2. Do earthquakes occur <u>just</u> along coastlines?   Yes   No

    If you chose "No," give an example of an area where earthquakes occur in the middle of an ocean, not on a coastline, and mark it on the map.

3. Do earthquakes occur <u>just</u> in hot climates?   Yes   No

    If you chose "No," give an example of a cold area that experiences earthquakes and mark it on the map.

4. Why do earthquakes occur where they do?

5. Two students are debating about why more earthquakes occur in California than in New York.

   Student 1: *Earthquakes occur where the faults are, so areas that have lots of faults also have lots of earthquakes. California has a lot of faults, so it has a lot of earthquakes, unlike New York.*

   Student 2: *That only partly answers the question because we need to know why faults occur where they do. Most faults occur along plate boundaries, so it's because California is on a plate boundary and New York is not.*

   Do you agree with one or both students? Why?

Part 2: Earthquakes and Plate Boundaries

The map to the right shows the locations of earthquakes in and near South America.

6. Based on the locations of the earthquakes, draw three or four lines on the map indicating where the major plate boundaries are located.

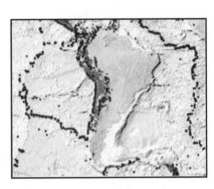

The map below shows the locations of plate boundaries in North America as solid black lines.

7. Based on the locations of the plate boundaries, what locations are least likely to experience earthquakes?

   A   B   C   D   E

8. Do all locations on the coast commonly experience earthquakes? Explain your answer using examples.

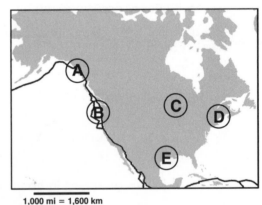

1,000 mi = 1,600 km

9. Do warm areas more commonly experience earthquakes than cold areas? Explain your answer using examples.

10. Explain to your friends why some places in North America get more earthquakes than others.

Lecture Tutorials for Introductory Geoscience

FAULTS

Below are cross-section diagrams of two faults in layered sedimentary rocks. The hanging wall is the side of the fault that is on top, and the foot wall is below the fault.

1. In the gray boxes ▢ in each diagram, label the hanging wall (HW) and foot wall (FW).

2. Along the fault in each diagram, draw arrows showing the direction of movement of each side based on the offset of the layers. ↙ or ↗

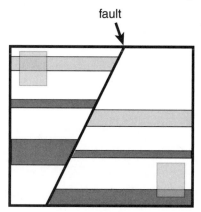

fault

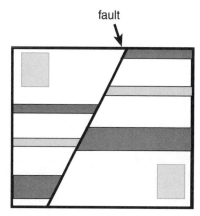

fault

Thrust/reverse fault: occurs in a region of compression where the hanging wall is pushed above the foot wall.

Normal fault: occurs in a region of tension where the hanging wall slides down relative to the foot wall.

3. Based on the direction of movement, label each of the diagrams above as "thrust/reverse fault" or "normal fault."

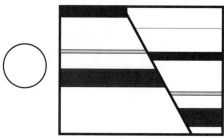

4. What type of fault is drawn?

   normal   or   thrust/reverse

6. What type of stress is shown?

   compression   or   tension

5. What type of fault is drawn?

   normal   or   thrust/reverse

7. What type of stress is shown?

   compression   or   tension

8. For each fault diagram above, draw one horizontal arrow in each circle ((→) or (←)) showing the stress direction.

9. On the map below, draw an arrow on each side of the two plate boundaries showing the direction of motion at each boundary.

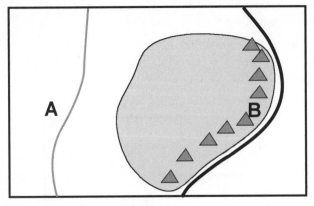

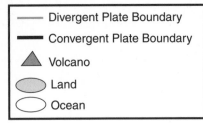

———	Divergent Plate Boundary
▬▬▬	Convergent Plate Boundary
▲	Volcano
⬭	Land
⬯	Ocean

10. Label Locations A and B on the map with "normal faults" or "thrust/reverse faults," depending on what you would expect based on the direction of stress.

11. Where would each of the faults below be formed on the map? Draw a line matching each fault with either Location A or B on the map above.

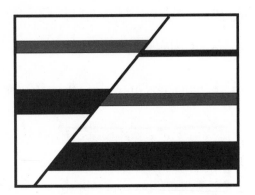

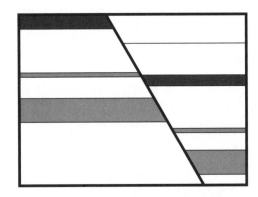

TSUNAMI

Part 1: Plate Boundaries and Tsunamis

Most tsunamis form when an earthquake occurs underwater. To create a tsunami, the earthquake needs to be large and move the seafloor up or down.

Below are cross-section diagrams showing the movement as a result of earthquakes that occur at each of the plate boundaries. The large black arrows show the movement of the plate.

1. Label each diagram of a fault with the correct plate boundary at which it normally occurs: divergent, convergent, or transform.

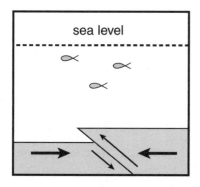 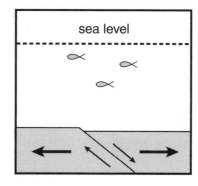 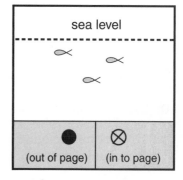

2. The largest earthquakes occur when rocks are compressed. Which type of plate boundary experiences the largest earthquakes?

   divergent   convergent   transform

3. At which type(s) of plate boundary do earthquakes cause the seafloor to move up and down?

   divergent   convergent   transform

4. Using the two criteria described above, which diagram(s) above illustrates where a tsunami usually forms?

   divergent   convergent   transform

5. Two students are debating about where tsunamis usually form.

   Student 1: *Tsunamis usually form only at convergent plate boundaries because that's the plate boundary where the largest earthquakes happen and the seafloor moves up or down during the earthquake.*

   Student 2: *I think tsunamis form at any plate boundary when any earthquake occurs underwater, as long as the earthquake is large enough. For example, California gets lots of earthquakes, and it is near the coast, so the San Andreas Fault, which is a transform fault, can cause a tsunami.*

   With which student do you agree? Why?

Part 2: Tsunami Formation

The sequence of diagrams to the right shows the formation of a tsunami along the edge of a continent.

6. What type of plate boundary is shown in the diagrams?

      convergent   divergent   transform

7. Why is there no movement between the plates at the circle before the earthquake?

8. What happens to the ocean water when the earthquake occurs?

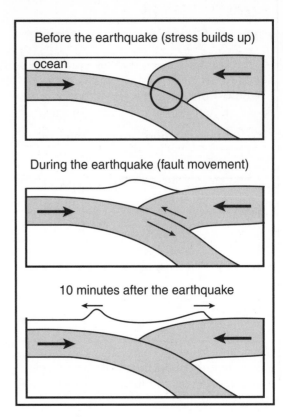

The map below shows the locations of plate boundaries. Recall that divergent boundaries tend to occur in the middle of oceans, whereas convergent boundaries tend to occur near the edges of oceans.

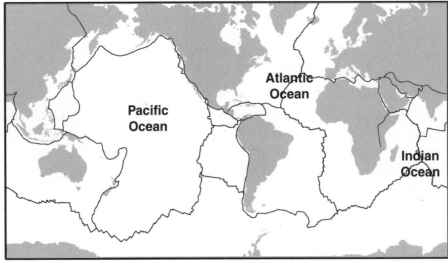

9. Which ocean has the most tsunami?   Atlantic   Pacific   Indian

   Explain your answer.

Lecture Tutorials for Introductory Geoscience

LANDSLIDES

Landslides are caused if the slope is too steep to support the weight of the land. Some of the factors that make landslides more likely include the following:

- steep surfaces
- water in the ground (it adds weight and decreases friction)
- lack of vegetation (roots stabilize the ground and remove water)
- planes of weakness in rock to allow sliding movement (such as parallel sedimentary layers, especially layers that include clay minerals)

1. Put an X next to the situations below that make landslides more likely.

_____ A hurricane saturates the ground with water.

_____ A wild fire burns plants and trees in the mountains.

_____ Plate tectonics cause mountains to get steeper over time.

_____ People water yards, causing water to soak into the ground.

Check your answers with another group.

In Questions 2–6, determine which house would be <u>least</u> likely to be destroyed by a landslide.

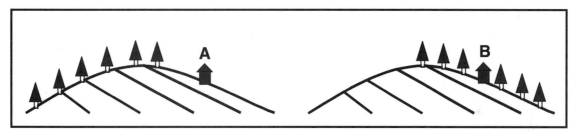

2. In which house would you prefer to live? A or B   Explain.

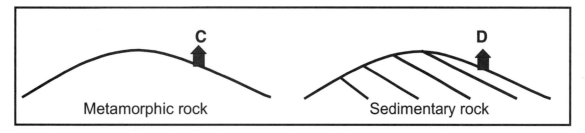

3. In which house would you prefer to live? C or D   Explain.

Lecture Tutorials for Introductory Geoscience

4. In which house would you prefer to live? **E** or **F**   Explain.

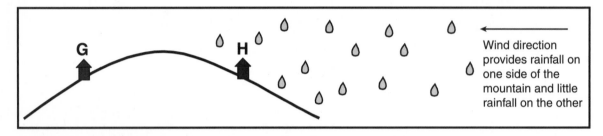

Wind direction provides rainfall on one side of the mountain and little rainfall on the other

5. In which house would you prefer to live? **G** or **H**   Explain.

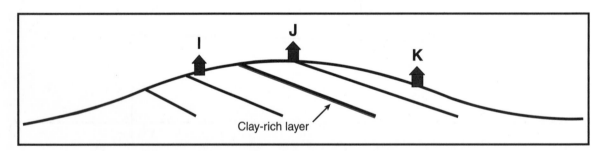

Clay-rich layer

6. In which house would you prefer to live? **I** or **J** or **K**   Explain.

7. After wildfires burn trees in mountains in California in the summer, there is often talk about the dangers of landslides during the winter wet season. a) Why are landslides more likely than before, and b) why would they happen in the winter instead of the summer?

In this diagram there are four houses. The bold surface is an impermeable paved slope.

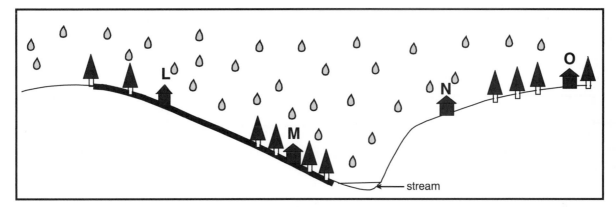

8. Which house is the <u>most likely</u> to be destroyed by a landslide? **L** or **M** or **N** or **O**

9. Explain the factors that you considered when deciding which house is most likely to be destroyed by landslides.

FLOOD FREQUENCY CURVE

Part 1: Constructing a Curve

A flood-frequency curve plots the discharge of a particular stream against how often that discharge occurs. In the chart below, the flood level of a stream in the United States was recorded each year between 1950 and 1999.

Year	Flood * discharge	Year	Flood * discharge	Year	Flood * discharge	Year	Flood * discharge	Year	Flood * discharge
1950	18	1960		1970	33, 18	1980		1990	50
1951	50	1961	18	1971		1981	18	1991	18
1952		1962	18, 33, 18	1972	33	1982		1992	
1953	18	1963		1973	50	1983	18	1993	
1954	33	1964	50, 18	1974	18	1984	33	1994	33, 18, 33
1955	18, 33	1965	33, 50	1975	18, 50	1985	18	1995	
1956		1966	18	1976	100	1986		1996	18
1957	18	1967		1977		1987	18, 18	1997	18
1958	33	1968	18	1978	33	1988		1998	
1959	18	1969		1979	18	1989	33	1999	18, 33

* in thousands of cubic feet per second (1000 ft^3/sec); some years have more than one flood

1. According to this table, which size flood happens more often?

    large   small

The table below summarizes the number of times each flood happens in 50 years (data are from the previous table). The recurrence interval indicates how often a flood of that size occurs.

Recurrence interval = 50 years ÷ the number of times that flood occurs in 50 years

2. Determine the recurrence interval (average number of years between floods) for the largest and smallest flood heights. The other recurrence intervals have been calculated for you.

Flood discharge	# of times in 50 years	Recurrence interval (1 flood every _____ years)
100,000 ft^3/sec	1	
50,000 ft^3/sec	6	8.3
33,000 ft^3/sec	13	3.8
18,000 ft^3/sec	25	

3. Does a flood with a large recurrence interval occur more or less often?   more   less

Below is the flood-frequency curve for this stream. The recurrence interval is plotted compared to the discharge (size of the flood).

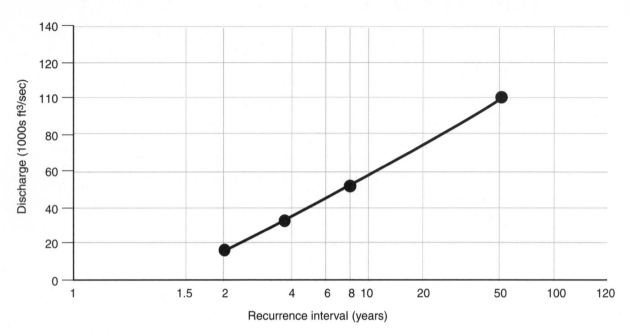

4. Large floods that are 75,000 ft³/sec occur on average once every _____ years.

5. Small floods that are 20,000 ft³/sec occur on average once every _____ years.

6. Predict the recurrence interval of an extremely large flood with a discharge of 140,000 ft³/sec.

Part 2: Predicting Floods

7. The chart makes it look like floods of a discharge rate of 50,000 ft³/s occur <u>exactly</u> once every eight years. Look at the first table. Are floods of certain sizes regularly spaced?

8. If there is a flood with a discharge rate of 50,000 ft³/s, will that flood happen again in exactly eight years? Explain your answer.

9. A flood has a recurrence interval of four years. This flood occurs in 2010. What is the chance the flood will happen in 2011?

10. Two students are thinking about how a flood in 2010 with a four-year recurrence interval affects the likelihood of a similar flood occurring in 2011.

    Student 1: *I think that it will not happen in 2011. It is just like an earthquake. If an earthquake just happened, then it will take many years for the stress to build up again for another earthquake. Since the flood just happened in 2010, it is not likely to happen the next year.*

    Student 2: *But weather this year doesn't care what happened last year. The flood has a 25% (1-in-4) chance of happening each year, so it would have a 25% chance of happening in 2011. Floods are not evenly spaced over the years.*

    Student 1: *I think it will happen in 2014, exactly four years later. The flood frequency is one flood every four years.*

    With which student do you agree? Why?

11. Find an example in the chart on the first page that shows that a large flood can happen two years in a row. Write the years you chose below.

12. A very large flood occurs in a small town. The local tourist board posts the message: "Come visit our city! We'll be safe from floods for another 100 years!" Do you agree with this message? What is the likelihood that the same flood will happen again next year?

FLOOD CURVES

Part 1: Rainfall

Look at the cross section below showing the ground surface and water table during a short, powerful rain storm.

1. What two possible things can happen to Raindrop A when it hits the ground?

2. Draw arrows showing the respective directions that surface water and ground water are flowing.

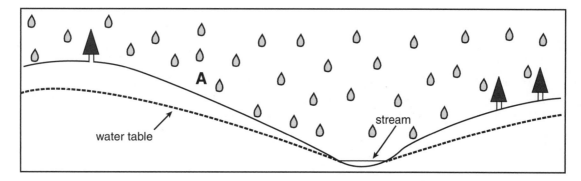

3. Where will Raindrop A eventually end up? _____

Part 2: Flood Curve

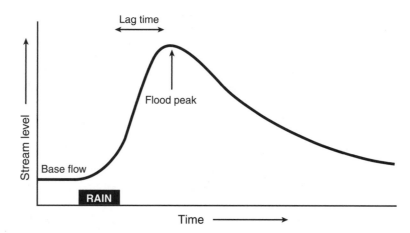

The graph above shows the level of water in the stream in Question 2. The black box indicates the duration of a short, powerful rain storm. Rain water reaches the stream by flowing along the ground surface and by soaking into the ground and traveling more slowly underground to the stream.

4. Is the flood at its highest during the worst of the rain storm?   Yes   No

5. What causes the peak of the flood to occur <u>after</u> the peak of the rain?

Lecture Tutorials for Introductory Geoscience

6. Imagine a scenario where no water soaks into the ground but instead runs quickly off the surface all at once into the stream. Predict how the following features of a flood would change, and explain how you came up with your answers.

    a) the length of the lag time   longer   shorter   because...

    b) the height of the flood peak   higher   lower   because...

    c) the duration of the flood   longer   shorter   because...

7. Two students are discussing how water soaking into the ground affects flooding.

    Student 1: *If water doesn't soak into the ground, it gets to the stream faster. And all the water would rush into the stream at once, so the flood height will be higher, but the flood will be short because all of the water arrives at nearly the same time.*

    Student 2: *I agree that the lag time will be shorter since the water reaches the stream quickly. But, I think that the higher the flood is, the longer it will last. More water will be in the stream for a longer period of time, increasing the total water in the stream.*

    With which student do you agree? Why?

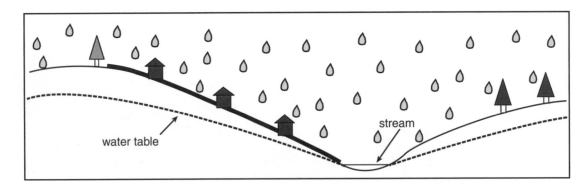

8. How would building a city with concrete parking lots and roads in a previously undeveloped area (see diagram above) affect the amount of water soaking into the ground versus running off the surface? Explain your answer.

9. Predict how building a city in a previously undeveloped area would affect (if at all) the following features of a flood:

a) the total amount of water in the flood   more   not affected   less

     briefly explain:

b) the length of the lag time   longer   not affected   shorter

     briefly explain:

c) the height of the flood peak   higher   not affected   lower

     briefly explain:

d) the duration of the flood   longer   not affected   shorter

     briefly explain:

10. On the diagrams below, the solid line is a flood curve for a rainstorm in a rural area. Which new curve (dashed) best represents how the flood curve would change if a city was built in the area?

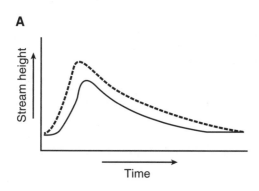

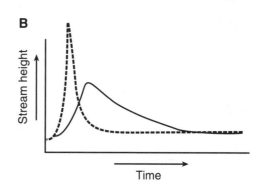

11. Check your answers to Questions 6, 9, and 10 to make sure they match.

12. Explain how people, even if they do not alter the stream, make floods higher by building in an area.

GROUNDWATER

Part 1: Pore Spaces

Pore spaces are the spaces between the pieces in a rock or sediment. Water that fills these pore spaces is called groundwater.

1. Imagine you pour water on sand. What happens to the water over time?

It sits on top of the sand.

It flows into the pore spaces between the sand grains.

It forms a small pool of water in the sand beneath the surface.

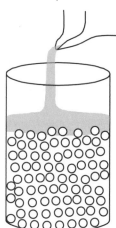 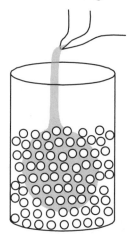 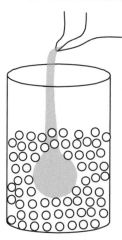

2. Imagine you pour water on the sedimentary rock sandstone. What happens to the water over time?

It sit on top of the sandstone.

It flows into the pore spaces between sand grains in the sandstone.

It forms a small pool of water in the sandstone beneath the surface.

3. Approximately how big are pore spaces in sediments or rocks?

Smaller than a grain of rice

About the size of a pea

About the size of an orange

Larger than a watermelon

4. Two students are discussing the best analogy for <u>most</u> groundwater.

Student 1: *I think most groundwater is like water in large underground holes or caves. When we look at water on Earth's surface, it gathers in big pools, so I think it would be the same underground. Besides, there is not enough room in rocks for water to fit.*

Student 2: *I think most groundwater is like water in a sponge. Water fills in lots of very small empty spaces within certain types of rocks, like sedimentary rocks. Caves do not exist in most places, so that can't be what most groundwater looks l ike.*

With which student do you agree? Why?

© W. H. Freeman and Company

Lecture Tutorials for Introductory Geoscience

Part 2: Aquifers

Aquifers are rock or sediment layers that contain groundwater and let groundwater move through them.

5. Porosity is the amount of a rock or sediment that is pore space. Would aquifers have high or low porosities? Explain your answer.

   high   low   because...

6. Permeability is the ability of liquid to pass through a material. Would aquifers have a high or low permeability? Explain your answer.

   high   low   because...

7. Circle the rocks listed below that would make a good aquifer.

   sandstone: high porosity, high permeability

   granite: low porosity, low permeability

   fractured limestone: high porosity, high permeability

   shale: high porosity, low permeability

   gneiss: low porosity, low permeability

8. Impermeable layers are layers that are barriers to groundwater flow. Put a star next to the three rocks above that would be impermeable.

9. Water gets added to groundwater by rain soaking into the ground. Circle the cross section below with the largest aquifer.

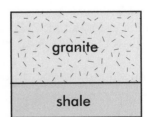

Lecture Tutorials for Introductory Geoscience

Oil also fills pore spaces in rock, like water does. After it forms deep beneath the surface, it slowly moves upward through the rock until it gets trapped by an impermeable layer. The cross section below shows layers of sedimentary rocks. The oil rig on the surface indicates where a well will be drilled.

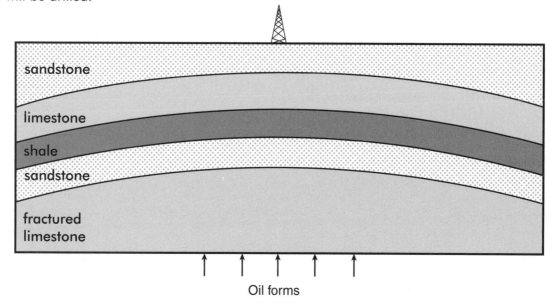

Oil forms

10. Draw arrows that continue the movement of the oil after it forms until the point it cannot move farther.

11. Draw in the well from the oil rig on the surface to the depth where you would drill for oil. Explain why you chose that particular depth.

WATER TABLE

The water table is the underground boundary between where the pore spaces are filled with air and where the pore spaces are filled with water.

1. To the right is a cross section of a cup of marbles partially filled with water. On the diagram, label the following areas:

   Pore space filled with water

   Pore space filled with air

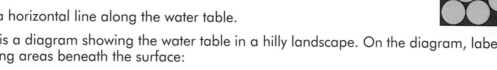

2. Draw a horizontal line along the water table.

3. Below is a diagram showing the water table in a hilly landscape. On the diagram, label the following areas beneath the surface:

   Pore space filled with water   Pore space filled with air

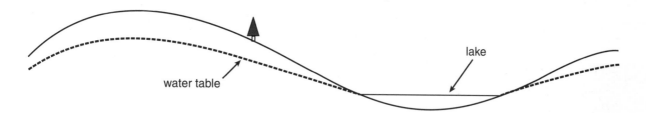

4. Two students are discussing the direction groundwater will flow if the water table is not flat.

   Student 1: *I think groundwater would flow so that the water table tries to flatten out due to gravity. So, water under the "hills" of the water table would flow downhill to fill in the "valleys."*

   Student 2: *I think that after water soaks down into the small pore spaces in the ground, it stops moving when it reaches the water table. There is no reason for it to flow sideways underground.*

   With which student do you agree? Why?

5. On the diagram above, draw three arrows showing the direction the ground water is flowing.

6. Is the groundwater flowing in the same direction in the entire area? Explain your answer.

7. Why is it that when you dig a hole in your backyard and try to fill it up with water, it drains into the ground, whereas the lake in the diagram above stays full? Use information about the water table and pore spaces to explain your answer.

Lecture Tutorials for Introductory Geoscience

GROUNDWATER CONTAMINATION

The contamination within groundwater flows along with the groundwater. As a result, by determining the flow of groundwater, you can determine the flow of the contamination.

1. The contours on the map to the right show the water table elevation above sea level.
   Why is Arrow C better than the other arrows at showing the direction groundwater will flow?

   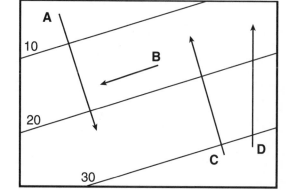

   Arrow A:

   Arrow B:

   Arrow D:

2. Which arrow shows the direction the contamination in the groundwater will flow?

   Arrow A   Arrow B   Arrow C   Arrow D

3. The contours on the map to the right show the water table elevation above sea level. The black triangle represents a septic tank. If this septic tank is improperly installed, which location would detect contamination first?

   E   F   G

   Explain your answer.

   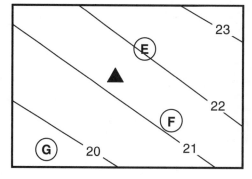

4. Three students are debating which location will first detect the contamination.

   Student 1: *I think that E will detect the contamination first because it is closest to the septic tank.*

   Student 2: *But you need to consider what direction the water will flow. I think that F will first detect the contamination because the contamination will flow along the lines down toward F.*

   Student 3: *The lowest elevation is toward G where the elevation is less than 20, so the water will flow from 21 to 20, and G will detect contamination first.*

   With which student do you agree? Why?

Lecture Tutorials for Introductory Geoscience

The figure below shows the water table elevation, three houses, three wells, and a septic tank.

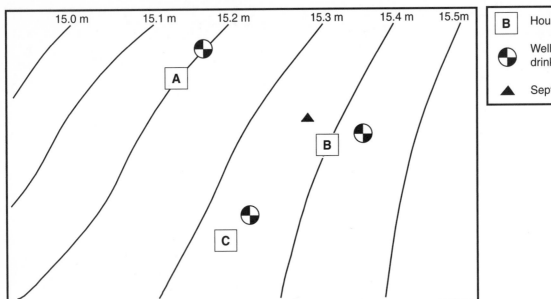

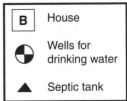

B	House
⊕	Wells for drinking water
▲	Septic tank

5. Draw at least three arrows on the map showing the direction of groundwater movement. Keep in mind your answer to Question 1 when drawing the arrows.

6. The triangle is a proposed location of septic tank for House B. Septic tanks may release contaminants in the water if constructed without the proper considerations. Draw an arrow showing the direction of the contaminant plume which may be released from the septic tank.

7. The wells of which houses, if any, could be affected by the septic tank?   A   B   C

8. Put a star on the map where you think would be a better place for a septic tank for House B. The tank, to be useful to the inhabitants of the house, needs to be located near the house. Explain your decision in the space below.

GLACIER MOVEMENT

A glacier is a mass of ice on land that flows under its own weight due to gravity.

1. Which way does a glacier flow?

    downhill   uphill   depends on the temperature

The glacier shown below is flowing downhill due to gravity. A spike was put in the glacier at the arrow.

Original glacier
before shrinking

2. On the diagram above showing the original glacier position, draw two arrows showing the direction that the glacier is moving.

Imagine the glacier shown is now shrinking due to warming temperatures.

3. Which diagram shows where the spike will be found in several years?

    A   B

    Explain your choice.

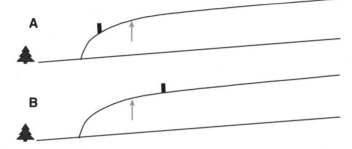

A

B

4. Two students are discussing which way glacier ice moves when a glacier is shrinking.

    Student 1: *Glacier ice always flows downhill because of gravity, and A shows the ice flowing downhill. The glacier shrinks because it melts faster than the glacier is flowing.*

    Student 2: *I think the ice would flow backwards and carry the spike with it, as shown by B. Because the glacier is shrinking, the ice would flow uphill.*

    With which student do you agree? Why?

5. When a glacier is getting smaller, it is often said that it is "retreating." Explain why this term may cause confusion in understanding how a glacier moves.

GLACIER BUDGET

The amount of ice in a glacier is a balance between additions (e.g., snowfall) and subtractions (e.g., melting).

1. On the cross-section of the glacier below, circle where most material leaves the glacier due to melting.

2. On the cross-section of the glacier below, put two stars where most material is added to the glacier due to snowfall that does not melt in the summer.

3. On the cross-section of the glacier below, draw an arrow showing which way the ice is moving.

4. What would happen to the size of a glacier if additions are more than subtractions?

   It would grow.   It would shrink.   It would stay the same.

5. What would happen to the size of a glacier if additions are less than subtractions?

   It would grow.   It would shrink.   It would stay the same.

6. What would happen to the size of a glacier if additions are the same as subtractions?

   It would grow.   It would shrink.   It would stay the same.

7. Think about a glacier where additions are less than subtractions. What would happen to the location of the bottom end of the glacier?

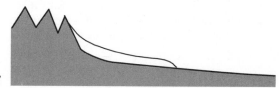

   It would advance (be more downhill).   It would retreat (be more uphill).

Remember that the ice itself does not move uphill. Instead, the ice is melting at the end of the glacier faster than it can be replaced by the ice flowing downhill.

8. For each row below, circle the situations that could cause the additions to be less than subtractions.

   increasing snowfall   decreasing snowfall

   increasing melting   decreasing melting

   increasing temperatures   decreasing temperatures

Lecture Tutorials for Introductory Geoscience

9. Two students are discussing why a glacier might grow.

    Student 1: *I think it would grow because it is getting colder. For example, this happened during the Ice Age—the world got colder and the glaciers got bigger.*

    Student 2: *That is only one reason that glaciers might grow. They might also grow if it snows more, which means an increase in the additions to a glacier.*

    With which student do you agree? Why?

On average, most glaciers around the world are currently shrinking because of warming average global temperatures.

10. What are at least two reasons why a few glaciers are growing?

11. Some people have argued that these growing glaciers show that the average global temperatures are not warming up. Based on your previous answers, explain why glaciers can grow—even if temperatures are getting warmer.

Lecture Tutorials for Introductory Geoscience

LONGSHORE CURRENT

The longshore current transports sediment parallel to, or along, the shore. When the current slows down, the sediment it is carrying is deposited, and when the current speeds up, more sediment is picked up (or eroded).

1. On the map view below, there is a structure, such as a groin, built into the ocean to block the longshore current. Circle the area where the longshore current slows down as a result of the structure.

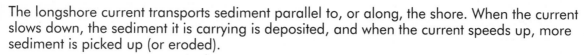

2. Based on the previous question, circle the diagram below that best shows where sediment will be deposited in response to building the groin.

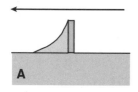 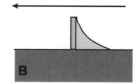 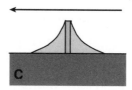 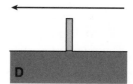

Explain your answer.

3. After the longshore current is disrupted, it continues once again and picks up and transports sediments. On the map view below, circle the area where the longshore current speeds up again after slowing down for the groin.

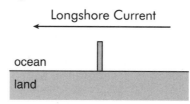

4. Based on the previous question, circle the diagram below that best shows where sediment will be eroded in response to building the groin.

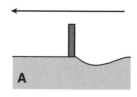 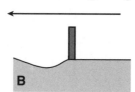 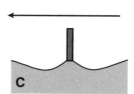 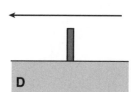

Explain your answer.

Lecture Tutorials for Introductory Geoscience

5. On the diagram below, sketch how the overall shape of the coastline will change in response to building the groin. Label your sketch with explanations of why the sediment is eroding in one place and depositing in another.

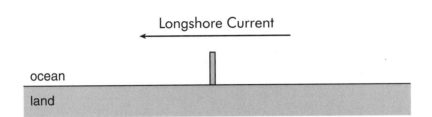

Longshore Current

ocean

land

6. On the map to the right, a sandy beach is shown in light gray. The rectangle represents a groin. Draw an arrow indicating the direction of the longshore current. Explain below how you determined the direction of the arrow.

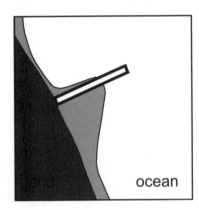

ocean

7. How does the total amount of sediment along a beach change when a groin is built to block the longshore current?

    increases   decreases   stays the same but is rearranged

Explain your answer.

Lecture Tutorials for Introductory Geoscience

TIDES

Part 1: Earth, Moon, and Sun Relationships

1. How long does it take for Earth to rotate on its axis one time?

    1 day   1 month   1 year

2. How long does it take for the Moon to travel around Earth one time?

    1 day   1 month   1 year

3. How long does it take for Earth and the Moon to travel around the Sun one time?

    1 day   1 month   1 year

4. Why does the Moon appear to travel across the sky in 12 hours (just as the Sun travels across the sky in 12 hours)?

High tides occur on Earth on the sides closest to the Moon and farthest from the Moon. Below is a figure showing Earth and the Moon as viewed from the North Pole. The dashed line on the diagram below shows tides on Earth as a result of the Moon (not to scale).

5. Label the high tides and low tides on the diagram.

6. At any point in time, how many high tides are on Earth?

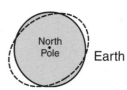

7. Two students are debating about why the tide goes in and out, alternating between high and low tide twice a day.

    Student 1: *I think that the tides change each day because Earth spins on its axis beneath the Moon, so a point will go from high to low tide as Earth rotates.*

    Student 2: *No, I think the tides change each day because the Moon moves around Earth. As the Moon moves around Earth, the high tides and the low tides follow the Moon.*

    With which student do you agree? Why?

Part 2: Semidiurnal Tide Cycle

Semidiurnal tides are two high tides and two low tides in approximately 24 hours.
Below are a set of figures showing Earth and the Moon over 24 hours as viewed from the North Pole. The star represents one place on the equator of Earth. The diagram is not to scale.

8. Draw the tidal highs (bulges) and lows on Earth created by the Moon.

9. Indicate if the star on the surface of Earth has a high tide or low tide at each point in time.

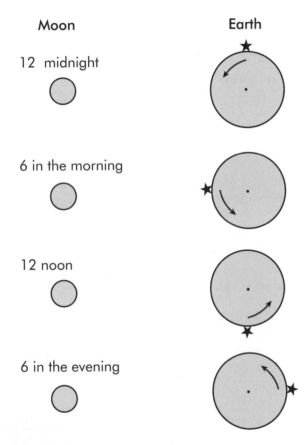

Moon   Earth

12 midnight

6 in the morning

12 noon

6 in the evening

10. How many hours separate a high tide from a low tide? _____

11. How many hours apart are high tides? _____

12. Planet Z is discovered. Its day is 10 hours long. Its moon revolves around it in 200 hours. How many hours apart are high tides on Planet Z? Explain how you determined your answer.

13. Summarize (in words or in a clearly labeled diagram) why semidiurnal tides occur and why there is a 24-hour cycle on Earth.

SPRING AND NEAP TIDES

Part 1: Earth, Moon, and Sun Relationships

It takes approximately four weeks (one month) for the Moon to travel around Earth. Below is a diagram showing the Earth, Sun, and Moon as viewed from the North Pole. The diagram is not to scale.

1. Draw an arrow on Earth to show Earth's rotation. How long does this take?

2. Draw an arrow on the Moon showing the direction of the Moon's motion around Earth. How long does this take?

Half of the Moon is shaded, and the other half is bright, showing the side that is lit from the Sun.

3. Circle the diagram to the right that shows what the Moon looks like as viewed from Earth.

4. What phase of the Moon is Question 3?   full moon   quarter moon   new moon

5. Draw in the tides. With a thin line, draw the tides on Earth caused by the Moon. With a dashed line, draw the tides on Earth caused by the Sun. Remember that the tides caused by the Sun are smaller than those caused by the Moon.

Part 2: Spring and Neap Tides

If the high tides from the Sun and Moon are in the same location, they add up, making them extreme (spring tides). If they are in different locations, they subtract from each other (neap tides).

6. In the configuration above, are the tides on Earth extreme or not extreme?

7. What is the name of this kind of tide?   spring tide   neap tide

---

Lecture Tutorials for Introductory Geoscience

8. Draw an Earth-Sun-Moon alignment below during a <u>spring tide</u>. There is more than one correct answer; however, there are incorrect answers. Remember that the Moon is always closer to Earth than the Sun ever is.

Shade half of the Moon, showing which side is lit by the Sun.

9. Circle the diagram to the right that shows what the Moon looks like as viewed from Earth.

10. What phase of the Moon is Question 9?   full moon   quarter moon   new moon

11. During what two phases of the Moon do spring tides occur?

   full moon   quarter moon   new moon

12. How long would you be able to see each phase?   one day   one week   one month

13. How many spring tides occur every time the Moon travels around Earth once?

14. Planet Z is discovered. Its day is 10 hours long. Its moon revolves around it in 200 hours. How many hours apart are spring tides on Planet Z? Explain how you determined your answer.

15. Explain why during some weeks tides are higher than during other weeks over the course of a month.

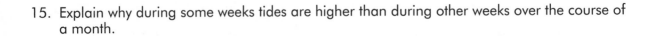

Lecture Tutorials for Introductory Geoscience

16. During a spring tide, explain why there is a pattern of two high tides and two low tides each day.

Please keep in mind that Question 16 is not asking why spring tides occur, so your answer does not need to mention the tides caused by the Sun.

Lecture Tutorials for Introductory Geoscience

CLIMATE CHANGE

CLIMATE CHANGE AND CARBON DIOXIDE

Part 1: Carbon Dioxide in the Atmosphere

Compare the chart to the right showing the amounts of different elements in the atmosphere with the graphs below showing the carbon dioxide concentrations and temperature anomalies over the last 250 years.

Current Atmosphere Composition of Select Elements	
Nitrogen	78.1%
Oxygen	20.9%
Carbon dioxide	0.038%

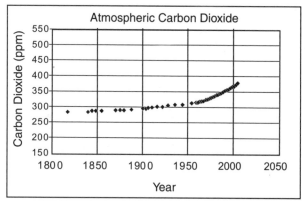

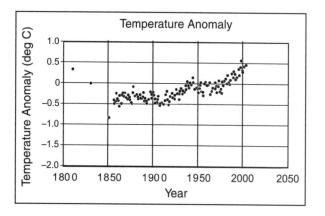

Data from the Carbon Dioxide Information Analysis Center

1. Look at the left graph. What is the carbon dioxide concentration today? _____ ppm

2. What is the trend in carbon dioxide concentration?   increasing   decreasing   constant

Use the trend in the graph to estimate the concentration of carbon dioxide in the atmosphere in 10 years and 100 years. People begin having headaches when carbon dioxide levels reach around 0.5% (5,000 ppm) and lose consciousness when levels reach 10% (100,000 ppm).

3. Will people be able to breathe in 10 years?   Yes   No

4. Will people be able to breathe in 100 years?   Yes   No

A temperature anomaly is how much the temperature is warmer or colder than normal. Note that it is measured in degrees Celsius.

5. Look at the graph on the right. What is the temperature anomaly today? _____

6. Therefore, is today warmer or colder than normal?   warmer   colder

7. What is the trend in temperature anomaly?   increasing   decreasing   constant

8. Based on the graph, what will the temperature anomaly be in 50 years? _____

9. What is the relationship between carbon dioxide and temperature anomaly?

Lecture Tutorials for Introductory Geoscience

Part 2: Comparison of Today to the Past

Below are graphs showing the levels of atmospheric carbon dioxide and the temperature anomaly for the past 400,000 years.

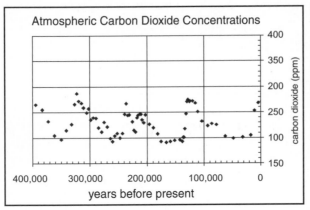

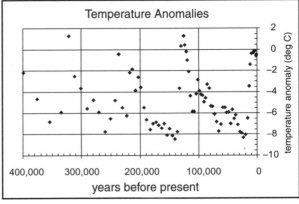

10. Use your answer to Question 1 to plot a point on the left graph indicating the carbon dioxide concentration today (note the different scale). How does the carbon dioxide concentration today compare to the concentrations in the past?

11. Use your answer for Question 5 to plot the current temperature anomaly on the graph above (note the different scale). How does the temperature anomaly today compare to the anomalies in the past?

12. Think about how hot it would need to be before humans could not survive. Did the temperature in the last 400,000 years get so hot that humans could not survive?   Yes   No

13. Based on past data, do you think the level of carbon dioxide in the atmosphere is going to increase to the point that Earth will warm up so much in the next 500 years that people will not be able to survive? Why or why not?

14. If humans do not need to worry about breathing or heat stroke issues, brainstorm some of the reasons why we are concerned about increasing levels of carbon dioxide in the atmosphere.

Lecture Tutorials for Introductory Geoscience

HOW THE GREENHOUSE EFFECT WORKS

The diagrams of two planets show how greenhouse gasses in the atmosphere cause a planet to be warmer than without them. Visible sunlight (black arrows) heats the surface. That heat radiates back into the atmosphere (gray arrows). Greenhouse gas molecules temporarily absorb the heat before radiating it again.

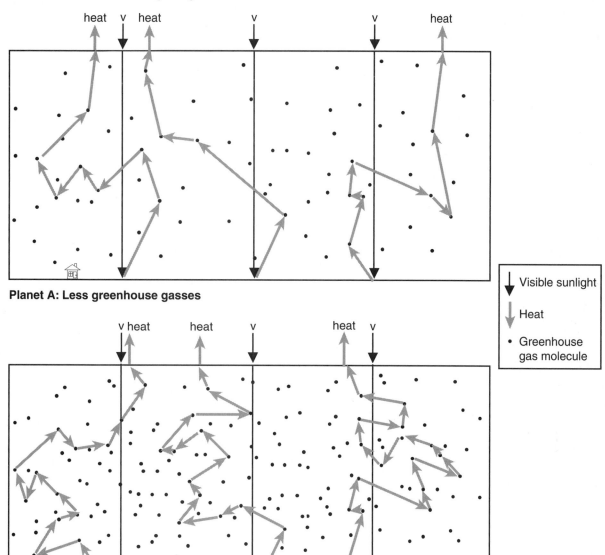

Planet A: Less greenhouse gasses

Planet B: More greenhouse gasses

Use the planets above with different amounts of greenhouse gasses to answer the questions. Note that the greenhouse gas molecules are drawn much larger than they are in reality.

1. Is visible sunlight (v ↓) affected by greenhouse gasses (•)?   Yes   No

2. Is heat energy (heat ↑) affected by greenhouse gasses (•)?   Yes   No

Lecture Tutorials for Introductory Geoscience

Circle how each of the following compares for the two planets.

3. The amount of visible sunlight (v ↓) entering the atmosphere.

    Planet A has more.   Planet B has more.   Both planets are the same.

4. The amount of heat energy (heat ↑) leaving the atmosphere.

    Planet A has more.   Planet B has more.   Both planets are the same.

5. The amount of time the heat energy (heat ↑) stays in the atmosphere.

    Planet A is longer.   Planet B is longer.   Both planets are the same.

6. The temperature of the atmosphere relates to how long heat energy stays in the atmosphere. Which scenario has a warmer atmosphere?

    Planet A is warmer.   Planet B is warmer.   Both planets are the same.

7. Count the arrows related to Questions 4 and 5. Change your answers above if necessary.

8. Below are three <u>incorrect</u> student descriptions on how the greenhouse effect works. For each statement, explain why it is incorrect.

Student A: Visible light from the Sun heats up the atmosphere of Earth.

Student B: The greenhouse effect is when greenhouse gasses form a hole in the atmosphere so more light can get in.

Student C: Greenhouse gasses magnify the Sun's heat that reaches the surface.

9. Ninety-seven percent of Venus' atmosphere is the greenhouse gas carbon dioxide. If Venus was the same distance from the Sun as Earth, explain how and why its average surface temperature would be different compared to Earth's average temperature.

THE GREENHOUSE EFFECT AND GLOBAL WARMING

The greenhouse effect is a natural process where heat energy interacts with greenhouse gasses in the atmosphere, causing warmer temperatures than there would be without greenhouse gasses. If there are greenhouse gasses in the atmosphere, then the greenhouse effect occurs.

Global warming is the increase of average temperatures around the world. The average temperature of the globe is warmer than before global warming began.

1. Fill in the blanks using "greenhouse effect" or "global warming."

   When the _____ is increased, that leads to _____.

   If greenhouse gasses are increasing, then _____ will occur.

Venus has an atmosphere made up of carbon dioxide, a greenhouse gas, and the levels of carbon dioxide in the atmosphere have not been observed to change.

2. Does Venus have a greenhouse effect?   Yes   No

   Does Venus have global warming?   Yes   No

   Explain your answers.

3. Two students are discussing whether or not humans cause Earth's greenhouse effect.

   Student 1: *Because the greenhouse effect is caused by gasses in the atmosphere, people cannot cause it. We can increase its strength by putting more gasses in the atmosphere, but the greenhouse effect is a natural process.*

   Student 2: *I think humans can cause the greenhouse effect. Because we are putting more greenhouse gasses in the atmosphere, we are causing the greenhouse effect to happen. Therefore, it is not a natural process, but instead it is caused by humans.*

   Student 1: *I think you are mixing up the terms global warming and greenhouse effect.*

   With which student do you agree? Why?

4. Match each phrase with either greenhouse effect or human-caused global warming (not both).

| greenhouse effect |

| global warming |

- a process that causes the temperature to be consistently warmer than it would be without greenhouse gasses
- always a natural process on a planet
- the increasing average temperatures of Earth's surface
- has been happening for all of Earth's history because there has always been carbon dioxide in the atmosphere
- currently caused by human civilization

You have a friend who says, "I've heard global warming has always happened because there has always been carbon dioxide in the atmosphere."

5. Explain how your friend is confusing the words global warming and greenhouse effect.

You have a friend who says, "This greenhouse effect thing is bad for Earth. We should take all carbon dioxide out of our atmosphere as fast as we can!"

6. Explain how your friend is misunderstanding the idea of Earth's greenhouse effect.

Lecture Tutorials for Introductory Geoscience

THE GREENHOUSE EFFECT AND THE OZONE HOLE

Below is a table showing properties of three types of light and how they react to different gasses in Earth's atmosphere and Earth's surface.

	Visible Light	IR Light (heat)	Ultraviolet Light (UV)
Ozone	passes through	passes through	absorbed by a thick layer of ozone
Greenhouse Gas	passes through	temporarily absorbed, heating up the atmosphere	passes through
Earth's Surface	absorbed and then radiated as IR	absorbed and radiated	absorbed

1. Circle two objects which we can see through Earth's atmosphere that illustrate visible light passes through both the ozone layer and the greenhouse gasses in Earth's atmosphere.

   a tree   the Sun   dirt   the Moon   clouds   a mountain peak

This diagram shows visible and UV light from the Sun, the ozone layer as a layer in the atmosphere (⠿), and greenhouse gasses throughout the atmosphere (▫).

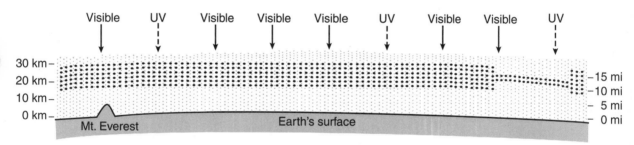

2. Label the location of the ozone "hole" on the diagram.

3. Where is visible light absorbed? (see chart at top) _____

4. On the diagram, continue drawing the visible light arrows until they are absorbed.

5. Where is UV light absorbed? (see chart at top) _____ _____

6. On the diagram, continue drawing the UV light arrows until they are absorbed.

7. Compare how IR light (heat) is affected by ozone versus greenhouse gasses.

Lecture Tutorials for Introductory Geoscience

When visible light hits the Earth's surface, it heats the surface and is radiated as IR light (heat). This heat energy is what heats up the Earth's atmosphere.

8. Where visible light hits Earth's surface, draw an arrow (↑) indicating it is radiated as IR (heat), and carefully draw the IR arrows until they are temporarily absorbed.

9. Does extra visible light hit Earth's surface because of the ozone hole?   Yes   No

10. Does the extra UV light that can come through the ozone hole affect the amount of IR light (heat) in the atmosphere?

     Yes   No

11. Two students are discussing how the temperature of Earth's surface is affected by a hole in the ozone layer.

     Student 1: I think an ozone hole affects the temperature because it lets in more UV light from the Sun, so the atmosphere and surface will be warmer.

     Student 2: But UV light doesn't heat up Earth's atmosphere or surface, so it won't actually warm them. So, the temperature is not affected by an ozone hole.

     With which student do you agree? Why?

12. Which one is most related to global warming, a larger hole in the ozone layer or the addition of extra greenhouse gasses, such as carbon dioxide? Explain why the other does not affect global temperature.

CONSEQUENCES OF GLOBAL WARMING

Global warming is the increase of average temperatures around the world. This increase can have serious consequences for the environment and people on Earth.

1. Below is a partial list of changes that could occur on Earth. Put an X in the correct box describing whether or not each is a result of global warming.

Not a Result of Global Warming	Current Consequence of Global Warming	Possible Changes
		melting glaciers
		lose land that is close to sea level due to rising sea level
		less sea-ice covering the oceans
		people will not have enough oxygen to breathe
		changing range of plants and animals that may cause extinctions
		humans cannot survive the consistently high temperatures
		more tsunami
		more severe hurricanes and tropical cyclones
		more extreme weather, causing floods and droughts
		bigger hole in the ozone layer

Earth's average temperature:
- today with a natural greenhouse effect: 14°C (57°F)
- if it had no greenhouse effect: –19°C (–2°F)
- in 100 years with global warming: ~16°C (~61°F)

2. Based on the temperatures above, what clothes, compared to today, would you need to buy if Earth had no greenhouse effect?

   much warmer clothing   about the same   much cooler clothing

3. Based on the temperatures above, what clothes will people, where you live, need to wear after human-caused global warming in 100 years?

   much warmer clothing   about the same   much cooler clothing

4. Based on your answers above, will the increasing temperatures caused by global warming cause people to die because of the constant, high temperatures?

   Yes   No

Lecture Tutorials for Introductory Geoscience

Carbon dioxide currently makes up 0.038% of our atmosphere (nitrogen is 78.1% and oxygen is 20.9%). Predictions are that in the next 100 years, human activity will cause the amount of carbon dioxide to double the preindustrial levels to 0.055%.

5. Will people have enough oxygen to breath, even if the carbon dioxide in the atmosphere doubles?

   Yes   No

Tsunamis usually occur when very large earthquakes suddenly displace a large amount of ocean water, causing a wave to form.

6. Will increasing average global temperatures by a few degrees affect how often a tsunami forms?

   Yes   No

Carbon dioxide does not cause the hole in the ozone. Rather, the hole in the ozone layer is caused by chemicals such as chlorofluorocarbons (CFC's). The hole in the ozone layer lets in extra UV light that causes sunburns. This UV light has no effect on the heat energy that causes the greenhouse effect.

7. Are the ozone hole and global warming related to each other in a significant way?

   Yes   No

8. After working through Questions 4–7, go back and check which four "X"'s you marked in the "Not a Result of Global Warming" column, and change them if necessary.

9. Two students are discussing the effects of global warming now and in the future.

   Student 1: *Most of the actual effects of global warming on the chart will happen in the future and will not really affect us.*

   Student 2: *I think most are happening now and will continue in the future, so we are being affected by them already.*

   With which student do you agree? Why?

10. Thinking about some of the effects of global warming, explain ways in which your city or state may be currently affected by global warming.

Lecture Tutorials for Introductory Geoscience

ALTERNATIVE ENERGY SOURCES

The world currently gets most of its energy from burning fossil fuels such as coal, oil, and natural gas. Burning fossil fuels releases the carbon contained in them, in addition to other pollutants, into the atmosphere. To address this concern, alternative energy sources are being used to produce energy.

hydroelectric dam   solar panels   wind turbine   nuclear power plant

1. Match each requirement on the right with a single alternative energy on the left.

hydropower

solar power

wind power

nuclear power

- large, fast-moving streams
- not too cloudy
- year-round streams
- Sun high in sky all year long
- place to build a large dam
- location for abundant solar panels
- location to dispose of radioactive waste
- constant wind
- people permitting nuclear reactions nearby
- visually nice location for windmills
- large source of water for cooling hot reactions
- very expensive

2. Which two alternative energy sources could be used by single houses or schools?

   hydropower   solar power   wind power   nuclear power

3. Which two alternative energy sources are best used by regions or states?

   hydropower   solar power   wind power   nuclear power

4. Three cities are looking to increase their use of alternative energy. Advise each city on which alternative energy would work the best in their situation, and explain your answer.

   City A is on the coast with a constant sea breeze. Many homeowners enjoy their pristine ocean view. The area is often cloudy with no major streams. The city will not be able to get funding approved to build a nuclear plant.

City B is frequently overcast with no consistent wind direction. It is near large mountain coastal streams, but these streams are important for fish migrations.

City C is not near any year-round streams because it is somewhat close to a sunny desert, home to many endangered species. Windy areas near the city are rough and mountainous.

5. Are any of the alternative energy sources without drawbacks? Give examples to support your answer.

6. Using the previous examples of cities, why is it so important to conserve energy?

Lecture Tutorials for Introductory Geoscience

HISTORICAL GEOLOGY

THE EARTH TIMELINE

1. Place the following five events in order on the lines below, from earliest to most recent:
   - Ice Age
   - dinosaur extinction
   - life evolved
   - Earth formed
   - fall of the Roman Empire

Earliest   Recent

_____   _____   _____   _____   _____

2. Label "Beginning of Earth" and "Today" on the correct sides of the timeline.

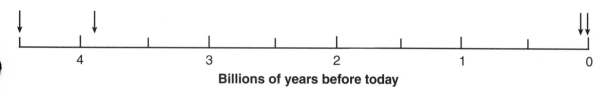

Billions of years before today

1 billion years = 1000 million years

3. Write the five events from Question 1 above to the corresponding arrows on the timeline. *Note:* Two of the events share an arrow because they occurred less than a million years apart.

The Ice Age occurred approximately 1 million years ago (0.001 billion years ago).

4. Does this date match the arrow you labeled as "Ice Age?"   Yes   No

   If it does not match, then change your labels.

Dinosaurs became extinct approximately 65 million years ago (0.065 billion years ago).

5. Does this date match the arrow you labeled as "Dinosaur extinction?"   Yes   No

   If it does not match, then change your labels.

6. Three students are discussing which two events share an arrow on the timeline.

   Student 1: *I think life evolved when Earth formed, so those two events happened at nearly the same time. Everything else happened much closer to today.*

   Student 2: *I think the Ice Age and the fall of the Roman Empire happened at nearly the same time because humans were alive for both of them.*

   Student 3: *I think the Ice Age caused the extinction of the dinosaurs, so those two events happened at nearly the same time.*

   With which student do you agree? Why?

7. Check your order for Question 1 and make sure it matches your answers to the other questions.

8. What two adjacent events had the most time between them? _____ _____

9. What two events happened in the last million years? _____ _____

10. People often consider dinosaurs to be one of the oldest forms of life on Earth. Based on your timeline and dates given, do you agree with this statement? Explain your answer.

11. Loch Ness is a lake that formed in Scotland during the Ice Age. Nessie, the "Loch Ness Monster," is said to live in the lake. One explanation people give for Nessie's existance is that she is a plesiosaur, a large marine reptile that went extinct with the dinosaurs. Based on your timeline and dates given, do you agree with this explanation? Explain your answer.

THE UNIVERSE TIMELINE

1. Place the following five events in order on the lines below, from earliest to most recent:
   - dinosaur extinction
   - Big Bang
   - Earth formed
   - Egyptian pyramids built
   - Ice Age

Earliest   Recent

_____   _____   _____   _____   _____

2. Label "Beginning of the Universe" and "Today" on the correct sides of the timeline below.

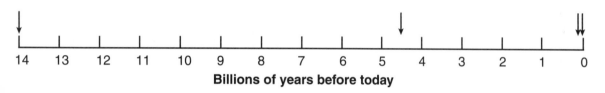

Billions of years before today

1 billion years = 1000 million years

3. Write the five events from Question 1 above the corresponding arrows on the timeline. *Note:* Two of the events share an arrow because they occurred less than a million years apart.

4. Two students are discussing which two events appear to have happened at the same time on the timeline.

   Student 1: *I think the Ice Age and Egyptian pyramids happened at nearly the same time because humans were alive for both of them. Everything else happened much longer ago.*

   Student 2: *I think the Big Bang formed Earth, so those two events happened at the same time. Everything else happened much closer to today.*

   With which student do you agree? Why?

5. Earth formed while the Solar System was forming. Add "Solar System formed" to the correct arrow on the timeline.

Lecture Tutorials for Introductory Geoscience

Dinosaurs went extinct approximately 65 million years ago (0.065 billion years ago).

6. Does this date match the arrow you labeled as "dinosaur extinction"?   Yes   No

   If it does not match, then change your labels.

7. Check your order for Question 1 and make sure it matches your answers to the other questions.

8. What two adjacent events had the most time between them? _____   _____

9. What two events happened in the last million years? _____   _____

10. Did the Big Bang immediately form the Solar System and Earth? Explain your answer.

UNCONFORMITIES

Part 1: Erosion

Not every place on Earth is a location where rocks are being deposited. Some places are where rocks are being eroded. The rock record is the rocks that remain that help us determine the history of an area.

1. In which of the following environments are rocks most likely to be eroded? This environment is an area where sediments are NOT deposited. Explain your reasoning.

    a. mouth of river

    b. ocean floor

    c. mountain

    d. coral reef

2. How could we tell if erosion happened in the rock record?

3. Two students are debating how erosion would be recorded in the rock record.

    Student 1: *If there was erosion, you wouldn't be able to see it because it gets rid of rocks. So there wouldn't be any record of it.*

    Student 2: *But, because rocks that were once there were taken away, you would see a missing surface at the top of the rocks. That missing surface would be visible in the rock record.*

    With which student do you agree? Why?

4. In the cross section to the right, what are <u>two</u> things that happened between the deposition of Sediment A and Sediment D?

    1.

    2.

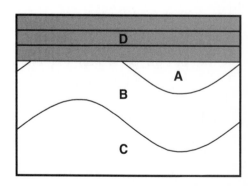

Lecture Tutorials for Introductory Geoscience

Part 2: Unconformities

In the cross section in Question 4, the surface between A/B and D is called an unconformity. An unconformity in the rock record is typically a surface where there was erosion of the rocks.

5. In which of the following environments is an unconformity most likely to form? Explain your reasoning.
    a. mouth of river
    b. ocean floor
    c. mountain
    d. coral reef

6. Why should your answers to Question 1 and Question 5 be the same? Change your answers as necessary.

Use the cross section to the right to answer Questions 7–9.

7. Draw a line along the unconformity in the cross section.

8. How do you know there is an unconformity in the cross section?

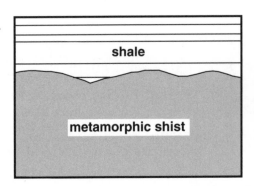

9. Describe three things that happened in the area to form the rocks in the cross section.

10. A person says to you, "I don't believe that Earth is thousands of millions of years old because there should be more sedimentary rock layers if Earth is that old." What would you say to that person to tell them that their argument does not make sense?

HOW DO WE KNOW WHEN DINOSAURS LIVED?

Part 1: Absolute Age Dating

Absolute ages are ages in years before present (e.g., This rock formed 100 million years ago (100 Ma)). Absolute ages are determined by studying the decay of unstable elements and knowing their half-lives. The table below summarizes characteristics of different unstable elements used to determine absolute geologic ages.

Unstable Element System	Used for...	Age Limits
$^{14}C \rightarrow {}^{14}N$ (carbon-14)	things that were once alive that retain their original carbon molecules	100 to 50,000 yrs before present (0.05 Ma)
$^{40}K \rightarrow {}^{40}Ar$ (potassium-argon)	volcanic rocks	0.05 Ma to 4600 Ma
$^{238}U \rightarrow {}^{206}Pb$; $^{235}U \rightarrow {}^{207}Pb$ (uranium-lead)	volcanic rocks	10 Ma to 4600 Ma

1. What system, if any, can be used to determine the absolute age of dinosaur bones?

2. Three students are debating how to date dinosaur bones.

    Student 1: *We can use carbon-14 because it's used to date things that were once alive. The flesh of a dinosaur was composed of carbon.*

    Student 2: *But dinosaurs lived more than 65 million years ago, so carbon-14 doesn't work. To date something that old, you need to use potassium-argon or uranium-lead.*

    Student 3: *Carbon-14 can't date things as old as dinosaurs and potassium-argon and uranium-lead only work for volcanic rocks, so none of these works.*

    With which student do you agree? Why?

Part 2: Relative Age Dating

Relative dating gives the ages of rocks relative to the ages of other rocks (e.g., This rock is older than that rock or fossil.). In undeformed sedimentary rocks, the oldest rock layer is found on the bottom, and the youngest rock layer is found on the top.

3. The figure to the right is a cross section through sedimentary rock layers. Which layer is the oldest?

    A   B   C   D   E

4. How old (relative to the other layers) is the dinosaur bone?

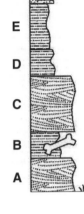

Part 3: Combining Dating Techniques

To the right is a cross section through sedimentary and volcanic rock layers. Sedimentary Rock Layer B contains a dinosaur bone.

5. Which two of the absolute age dating methods would you use to determine the age of volcanic Rock Layers A and C?

6. How could you determine the age of the dinosaur bone?

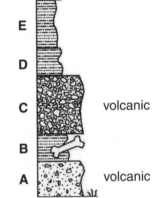

7. If volcanic Rock Layer A is determined to be 90 Ma, and volcanic Rock Layer C is determined to be 80 Ma, what is the age of the dinosaur bone?

8. Two students are debating the age of the dinosaur bone.

    Student 1: *The bone is between 80 and 90 million years old. You can't be more specific because you can't determine how fast the layers were deposited.*

    Student 2: *I think the dinosaur lived 85 million years ago because the bone is exactly in the middle of the two volcanic layers.*

    With which student do you agree? Why?

Lecture Tutorials for Introductory Geoscience

DETERMINING RELATIVE ROCK AGES

Part 1: Using the Principles of Relative Dating

For each cross-section diagram, you will number the units and events in order of formation. The oldest unit is number 1. Continue until the youngest unit is numbered.

1. Number the four events form oldest to youngest.

   Remember, 1 is the oldest.

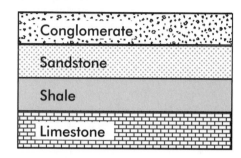

2.

a. Which is older, Unit A or Unit F?

b. Which is older, Unit C or the erosion surface (unconformity, shown by the wavy line)?

c. Number the seven events, including the unconformity, from oldest to youngest.

3.

a. What is the diagonal straight line through the cross section? _____

b. Which unit is a pluton (intrusive igneous rock)? _____

c. Which is <u>older</u>, Unit R or the fault that cuts Unit R? _____

d. Number the six events, including the fault, from oldest to youngest.

Part 2: Correlating Rock Layers

The two cross sections below show rocks in Arizona and Utah. Fossils are indicated by the symbols in each section.

4. Determine the order of events, including unconformities, for each area individually. Write your answers on the lines given. All five lines will be used for each cross section.

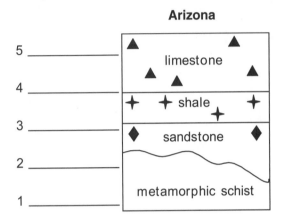
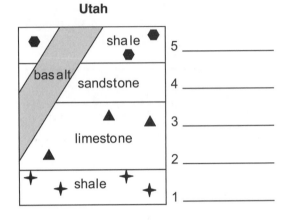

Arizona   **Utah**

5 _____

4 _____

3 _____

2 _____

1 _____

5 _____

4 _____

3 _____

2 _____

1 _____

5. Identify the units that have the same fossils in Arizona and Utah.

   Draw the fossils here:

6. Draw arrows on the diagram between the Arizona and Utah columns indicating which units are the same age because they have the same fossils.

7. If you went to Utah, what type of rock would you expect to find below the lower shale unit with the same fossils(✦)?

8. Determine the history of the entire area by correlating the two cross sections using the fossils present in the units. If there are units in different places with the same fossil, they were deposited at the same time and will be written only once in the combined column.

youngest

_____

_____

_____

_____

_____

_____

_____

oldest

Lecture Tutorials for Introductory Geoscience

INDEX FOSSILS

Index fossils are used to determine the relative ages of distant rocks. If the same index fossil is found in two rocks, then the sediments that formed the rocks were deposited at approximately the same time. To be a good index fossil, a species needs to 1) live for a short time range, 2) have a large area where it lived, and 3) be easily identifiable.

1. Based on <u>only</u> the area in which they live, circle the two animals that would make better index fossils.

| graptolite, floating throughout the ocean | coral, living only in clear, shallow ocean water | ammonite, swimming throughout the ocean | crab, living only on sandy, shallow ocean floor |

2. The diagram to the right illustrates the geologic time range of two brachiopods. Based on <u>only</u> the time period during which they lived, which brachiopod makes the better index fossil?

    Lingula brachiopod   *Mucrospirifer* brachiopod

    Explain your answer.

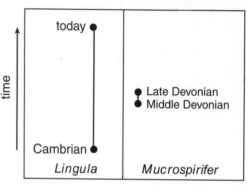

The diagrams below show fossils in rock columns from four different areas. The oldest rocks are at the bottom and the youngest are at the top. The columns do not necessarily represent the same length or period of time.

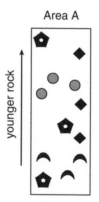   

3. Examine just Area A. Which fossil has the longest geologic time range?

4. Examine just Area A. Which two fossils have the shortest time range?

5. Examine all four areas. Which two fossils are widespread across all areas?

6. Examine all four areas. Which fossil would make the best index fossil?

    Explain your answer using the three characteristics for good index fossils.

Lecture Tutorials for Introductory Geoscience

HALF-LIFE

Part 1: A Magic Half-Life Candy Shop

A candy shop puts 1000 magic green candies in a jar. After 10 years, half of the green candies turn red. After another 10 years, half of the remaining green candies turn red. This process continues until all the green candies turn red.

1. After 10 years, how many candies remain green? _____ How many are red? _____

2. Fill in the table below and graph the results on the chart.

Number of Years	Number of Green Candies
0	1000
10	
20	
30	
40	
50	

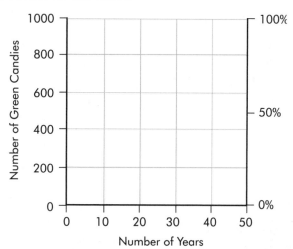

3. When you go to the shop and look in the magic candy jar, 125 candies are green. How many years have the candies been in the jar? Explain how you know, or write down any equations you used.

4. How many years does it take for 75% of the candies to turn red (25% of the candies remain green)?

The half-life of the magic candies is how long it takes for half of them to change colors.

5. What is the half-life of the candies in this jar?

6. In a different candy jar, half of the yellow candies change to blue every 47 years. What is the half-life of the candies in this other jar?

Part 2: Radioactive Elements

Radioactive elements work in a similar way as the candies in the magic candy shop.

7. Match the phrases relating to the magic candy shop to the phrases relating to radioactive elements in minerals. Do this by drawing lines between the two columns.

    a. The green candies are put into a jar.

    b. All candies begin green.

    c. No candies disappear.

    d. Candies change from green to red.

    e. The half-life of the candy is the time it takes for half the candies to change color.

    1. All atoms begin as parent isotopes.

    2. Atoms change from the parent to the daughter isotope.

    3. The half-life of the radioactive element is the time it takes for half the parent isotope to change to the daughter isotope.

    4. No radioactive atoms disappear into nothing.

    5. The parent atoms form in a crystal.

To determine the age of a rock (just like determining the time the magic candies have spent in the jar) you need to know two things: 1) how long the half-life is, and 2) how much of the original parent is remaining.

8. Geologists can measure the half-life of radioactive elements in a lab, so we know the half-life for common radioactive elements. Therefore, to determine the age of a particular rock, what would you need to measure in that rock?

A geologist is studying two different rhyolite flows to determine if they were erupted at the same time.

9. Rhyolite #1 has 50% of the Parent Isotope F remaining. How many half-lives have occurred?

10. Rhyolite #2 has 75% Daughter G and 25% Parent H. How many half-lives has the rock gone through?

11. Parent Isotope F in Question 9 has a half-life of 100 million years. Use your answer to Question 9 to determine the age of Rhyolite #1.

12. The half-life of Parent H-Daughter G in Question 10 is 200 million years. Use your answer to Question 10 to determine the age of Rhyolite #2.

13. Are the two rhyolite flows the same age?   Yes   No

One-hundred kilometers (60 miles) from Rhyolite #1 and Rhyolite #2, the geologist found an ash layer in the rocks. In order to determine how explosive the volcano could be, the geologist wanted to know if the ash came from the same eruption as either of the rhyolites.

14. A different parent-daughter pair in the felsic ash has a half-life of 50 million years. There is 25% of this parent still remaining. How old is the felsic ash?

15. Another parent-daughter pair is measured in the felsic ash. This pair has a half-life of 25 million years. There is 1/16 of the parent remaining. How old is the felsic ash?

16. The felsic ash is the same age as which rhyolite?   Rhyolite #1   Rhyolite #2

Lecture Tutorials for Introductory Geoscience

SCIENTIFIC HYPOTHESES OF DINOSAUR EXTINCTION

Part 1: Hypotheses

A scientific hypothesis needs to

1. be supported by the majority of current data.

2. be testable.

An alien on Earth is wondering why a rubber ball falls back down to the ground after it is thrown into the air. It comes up with several ideas about the ball.

a. Gravity is pulling the ball to the ground.

b. A mystical force that cannot be measured is pushing the ball down.

c. Earth's magnetic field is pulling on the rubber ball.

1. Which statement is **NOT** a hypothesis because it is not <u>testable</u>?   a   b   c

2. Which statement is **NOT** a hypothesis because it is not <u>supported</u> by current data?   a   b   c

3. Which statement **IS** a scientific hypothesis?   a   b   c

Part 2: Dinosaur Extinction

Below are possible scenarios explaining the extinction of the dinosaurs.

a. Dinosaurs were killed off by a virus.

b. A large meteorite impact caused the climate to change so some plants and animals could no longer survive.

c. Volcanic eruptions caused the climate to change so some plants and animals could no longer survive.

d. Mammals ate all the dinosaur eggs.

4. Determine if each statement above is a valid hypothesis. Be sure to explain your answer.

a.   Yes   No   because...

b.   Yes   No   because...

c.   Yes   No   because...

d.   Yes   No   because...

5. Two students are debating the hypotheses of dinosaur extinctions.

   Student 1: *I think that the meteorite and volcano statements are valid hypotheses, but the other two are not. You can't test the fossil record to find out if they are true, and they don't explain animals other than dinosaurs going extinct.*

   Student 2: *I think that all of the statements are all valid scientific hypotheses explaining dinosaur extinctions. I saw all of them on a dinosaur program on TV, and they all seem possible. No person was there to watch the dinosaur extinction, so all of the scenarios are hypotheses.*

   With which student do you agree? Why?

CONVERGENT AND DIVERGENT EVOLUTION

Convergent evolution is when <u>distantly related</u> organisms evolve to be more <u>similar</u> to each other because they live in the same ecologic niche.

Divergent evolution is when <u>closely related</u> organisms evolve to be more <u>different</u> because they live in different ecologic niches.

Part 1: Characteristics and Ecologic Niches

1. Flying animals, such as birds, insects, and bats, occupy the same ecologic niche. Name one physical characteristic that these animals share.

2. Predators swimming in oceans, such as sharks, fish, and seals, occupy the same ecologic niche. Name one physical characteristic that these animals share.

3. Why would organisms occupying the same ecologic niche evolve to be similar to each other?

Part 2: Whales

Whales are descended from (and therefore closely related to) land-living mammals, similar to hippos. As they began living in the ocean millions of years ago, whale ancestors that had variations that gave them advantages for living in oceans were better able to live and reproduce, passing along those variations, slowly making whales better adapted to oceans. Sharks have gills to remove oxygen from the water and are cold-blooded. They have remained relatively unchanged during these millions of years.

whale

shark

4. To what animal classification do whales belong?

   fish   mammal   reptile   amphibian

5. To what animal classification do sharks belong?

   fish   mammal   reptile   amphibian

---

6. Are whales and sharks closely related?   Yes   No

7. Are whales and sharks overall similar in outward appearance compared to whales and most other mammals?

    Yes   No

8. Therefore, based on your responses to the last two questions, are whales and sharks examples of convergent or divergent evolution? Explain your answer.

Ichthyosaurs are reptiles that swam in the oceans during the time of the dinosaurs. They had a streamlined body, powerful tails and fins, and a long mouth full of sharp teeth.

ichthyosaur

9. For each pair of organisms, determine if they are examples of convergent or divergent evolution, and explain your answer.

    Whales and ichthyosaurs: convergent   or   divergent   because...

    Whales and hippos: convergent   or   divergent   because...

Part 3: Penguins

Penguins and ostriches are both birds. However, they live in different ecologic niches, since penguins swim in the ocean and eat fish, and ostriches live in grasslands and eat plants and insects.

penguin   ostrich

Lecture Tutorials for Introductory Geoscience

10. Are penguins and ostriches closely related compared to all living animals?   Yes   No

11. Are penguins and ostriches similar in outward appearance compared to all other birds?

    Yes   No

12. Therefore, are penguins and ostriches examples of convergent or divergent evolution? Explain your answer.

Eagles are predatory birds that have sharp beaks and talons to catch prey. Seals are mammals with streamlined bodies, powerful swimming capabilities, and thick insulation for cold water.

eagle

seal

13. For each pair of organisms, determine if they are examples of convergent or divergent evolution, and explain your answer.

    Eagles and penguins: convergent   or   divergent   because...

    Seals and penguins: convergent   or   divergent   because...

Lecture Tutorials for Introductory Geoscience

THE FUNCTION OF *STEGOSAURUS* PLATES

● Part 1: Evaluating Different Hypotheses

Three hypotheses of the function of *Stegosaurus* plates are:

    a) protection and defense.

    b) intraspecies display (display to other *Stegosaurus,* often related to mating behavior).

    c) thermoregulation (regulation of the dinosaur's temperature).

1. Which hypothesis do you initially think is more likely to be correct? You may choose more than one.

   Briefly explain your reasoning.

2. Because it is impossible to observe the behavior of a living *Stegosaurus,* describe two or three ways that paleontologists can learn about the function of *Stegosaurus* plates.

● Part 2: Evidence Supporting Different Hypotheses

Different lines of evidence of the function of *Stegosaurus* plates are given below. For each line of evidence, determine which hypothesis the evidence supports. The evidence can support more than one hypothesis.

3. The shapes and patterns of the plates and spines for different species of stegosaurs are nearly always unique to that species.

4. The plates are made up of bone that forms a hollow honeycomb pattern.

5. The outsides of the plates are covered with grooves that are likely to have been the locations of blood vessels. Blood would warm or cool depending on the location of the sun and wind as it flowed near the surface of the plates.

Lecture Tutorials for Introductory Geoscience

6. The large, flat surface of the plates would catch a lot of sunlight.

7. Fossils of juvenile *Stegosaurus* do not appear to have plates.

Part 3: Conclusion

8. Examine your answers in Part 2. Which hypothesis is best supported by the data? There may be more than one. Explain your answer.

GENERAL

WHAT DO GEOLOGISTS DO?

1. Below is a list of scientific questions. Put a check in front of the problems that geologists solve. Include questions that require geologists to team up with other scientists such as biologists or astronomers.

_____ What was the climate like a million years ago, and is it changing today?

_____ How often does a river flood, and what areas would be under water?

_____ What areas will be affected when a volcano next erupts?

_____ Where is the best place to drill for oil, and how much oil is available?

_____ Where does contamination go that is underground?

_____ Where are good places to go fossil hunting?

_____ Are glaciers shrinking, and why?

_____ How fast could dinosaurs run?

_____ Was there life on Mars?

2. Two students are discussing what geologists do.

   Student 1: *I think geologists study rocks all over the world. When they find rocks, geologists identify them and classify them based on their characteristics.*

   Student 2: *I think geologists study Earth around us. They are like detectives, trying to figure out what happened in the past and what will happen in the future.*

   With which student do you agree? Why?

3. Three students are discussing why they want to be geologists.

    Student 1: *I like to spend time outdoors and travel to interesting locations.*

    Student 2: *I want to do something where I can make lots of money.*

    Student 3: *I want to help people, especially those that could be affected by natural or environmental disasters.*

    All of these are valid reasons to become a geologist. Which student or students best express opinions similar to yours? Why?

4. Because geologists do everything on the list in Question 1, summarize how your life would be different without geologists.

Lecture Tutorials for Introductory Geoscience

DENSITY

Part 1: Density

Density is the amount of material in a given volume. It can be thought of as how compact something is. If two objects are the same size, but one is heavier, it is more dense.

1. For each of the three rows below, circle which is more dense.

<table>
<tr><td>a bucket of
ping pong balls</td><td>or</td><td>a bucket of
golf balls</td><td>or</td><td>They are the same density.</td></tr>
</table>

<table>
<tr><td></td><td>or</td><td></td><td>or</td><td>They are the same density</td></tr>
<tr><td>a small piece
of granite</td><td>or</td><td>a large piece
of granite</td><td>or</td><td>They are the same density.</td></tr>
</table>

2. Three students are discussing the density of different sized rocks.

   Student 1: *I think a small rock is less dense than a large one. When you pick them up, the small rock is much lighter than the big rock. Since density is the same thing as weight, a heavier rock is a denser rock.*

   Student 2: *I think things with the same density but different sizes can have different weights. The atoms in the two rocks are equally compact, so the density is the same.*

   Student 3: *I think it is easiest to compare things of the same size. If you broke a piece off the big rock that was the same size as the little rock, they would have the same weight, so that means the rocks have the same density*

   With which student or students do you agree? Why?

3. A boulder is more dense than water. What happens to a boulder if you throw it in water?

   It rises.   It sinks.   It depends on the size of the boulder.

4. Hot air is less dense than cold air. What happens to a balloon if you fill it with hot air?

   It rises.   It sinks.   It depends on the size of the balloon.

5. Circle the sentence that best summarizes the interaction between materials of different densities.

    More dense material sinks and less dense material rises.

    Less dense material sinks and more dense material rises.

    The interaction varies, depending on the size.

Part 2: Density in the Geosciences

Hot mantle rock is less dense than cooler mantle rock. Mantle rock gets heated deep within Earth and cools near Earth's surface. The diagram below uses large arrows to show the path rock takes in the mantle.

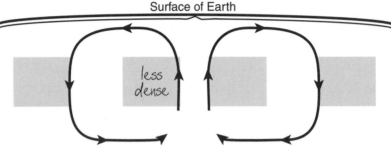

6. Use the words "hotter" and "cooler" to label the rocks Deep within Earth and the rocks at the Surface of Earth.

7. As shown, label the gray boxes "more dense" or "less dense" to explain why the mantle rock is moving up or down due to differences in density.

Density differences are also important when considering the lithosphere.

8. When ocean lithosphere and continental lithosphere move toward each other, the ocean lithosphere is subducted down beneath the continental lithosphere. Which is more dense?

    ocean lithosphere   continental lithosphere   the thicker lithosphere

9. As oceanic lithosphere gets older, it cools and becomes more dense. What would happen as one old and one young oceanic tectonic plate come together? Determine which plate in each diagram below is more dense, and then circle the correct diagram.

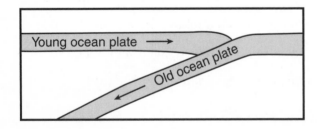 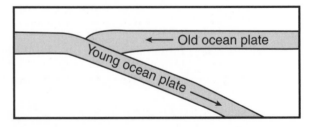

10. Explain your choices to Questions 8 and 9 in terms of how density affects the way two plates behave at a convergent boundary with subduction.

ISOSTACY

●Earth's surface (the lithosphere) sinks down if a large amount of weight is added to it. On the other hand, the surface rises if weight is taken off. This principle is called isostacy.

The series of diagrams below shows a growing and shrinking ice sheet on the lithosphere during the Ice Age to illustrate this concept. The movement of the lithosphere is exaggerated.

Before	Ice sheet grows	Ice sheet stable	Ice sheet melts	Back to before

1. Which two diagrams are isostatically unstable, so the ground is moving up or down?

    before   ice sheet grows   ice sheet stable   ice sheet melts   back to before

2. Draw an arrow showing the lithosphere moving up or down on the two unstable diagrams.

3. At the end of the most recent Ice Age, large ice sheets melted off of Canada and Northern Europe. Predict which way the ground is slowly moving in these locations currently and explain your answer.

    up   down   not moving   because...

4. Circle the way the <u>bottom</u> of the lithosphere would move in each of the following situations:

    deposition of thick layers of sediment   up   down
    large mountains formed   up   down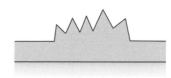
    erosion of mountains (removal of rock)   up   down

5. Two students are discussing the shape and thickness of the lithosphere in a mountain range.

    Student 1: *I think the lithosphere would be thicker beneath mountains because the weight of the high mountains would push down, like I drew to the right. Just like when the glacier pushed down on the lithosphere.*

    Student 2: *But mountains are when the lithosphere is pushed up at plate boundaries. So, I think that the bottom of the lithosphere would stay at the same depth and not change even with the weight of the entire mountain range on top of it, like my drawing to the right.*

    Use your knowledge about how the extra weight of glaciers affects the lithosphere to explain with which student you agree.

Lecture Tutorials for Introductory Geoscience

Mountain ranges erode away by rain and other elements.

6. On the second diagram above, draw an arrow showing the up or down movement of the bottom of the lithosphere after the mountains were eroded away.

7. Based on your arrow, draw the new location of the rocks that were at location "X".

8. Using the principle of isostacy, explain why rock formed deep within the lithosphere (such as the metamorphic rock at location "X") can be seen in eroding mountain ranges.

9. Using the principle of isostacy, explain why thick layers of sediment can be deposited in the ocean at the edge of the continent (like at a delta) without piling up above sea level.